The Aesthetics of Scientific Experiments

The relationship between aesthetics and science has begun to generate substantial interest. However, for the most part, the focus has been on the beauty of theories, and other aspects of scientific practice have been neglected. This book offers a novel perspective on aesthetics in experimentation via ten original essays from an interdisciplinary group comprised of philosophers, historians of science and art, and artists.

The collection provides an analysis of the concept of beauty in the evaluation of experiments. What properties do practising experimenters value? How have the aesthetic properties of scientific experiments changed over the years? Secondly, the volume looks at the role that aesthetic factors, including negative values such as ugliness, as well as experiences of the sublime and the profound, play in the construction of an experiment and its reception. Thirdly, the chapters provide in-depth historical case studies from the Royal Society, which also allows for a study of the depiction of scientific experiment in artworks, as well as contemporary examples from the Large Hadron Collider and cases of experiments designed by artificial intelligence. Finally, it offers an exploration of the commonalities between how we learn from experiments on the one hand and the cognitive value of artworks on the other.

The Aesthetics of Scientific Experiments will be of interest to researchers and advanced students working in philosophy and history of science, philosophy and history of art, as well as practising scientists and science communicators.

Milena Ivanova is a Byefellow at Fitzwilliam College, University of Cambridge, specialising in the aesthetics of science. Her project investigates the aesthetic values of scientific experiments. She is the author of *Duhem and Holism* (Cambridge University Press) and co-editor of *The Aesthetics of Science: Beauty, Imagination and Understanding* (Routledge).

Alice Murphy is a postdoctoral researcher at Ludwig Maximilian University of Munich (LMU), Munich. Her research focuses on aesthetic values in science, the epistemic role of imagination, and scientific thought experiments and models.

Routledge Studies in the Philosophy of Science

Logical Empiricism and the Physical Sciences
From Philosophy of Nature to Philosophy of Physics
Edited by Sebastian Lutz and Adam Tamas Tuboly

The Explanatory Autonomy of the Biological Sciences
Wei Fang

The Tools of Neuroscience Experiment
Philosophical and Scientific Perspectives
Edited by John Bickle, Carl F. Craver, and Ann-Sophie Barwich

The Internet and Philosophy of Science
Edited by Wenceslao J. Gonzalez

New Philosophical Perspectives on Scientific Progress
Edited by Yafeng Shan

Evidence Contestation
Dealing with Dissent in Knowledge Societies
Edited by Karin Zachmann, Mariacarla Gadebusch Bondio, Saana Jukola, and Olga Sparschuh

Conjunctive Explanations
Nature, Epistemology, and Psychology of Explanatory Multiplicity
Edited by Jonah N. Schupbach and David H. Glass

The Aesthetics of Scientific Experiments
Edited by Milena Ivanova and Alice Murphy

For more information about this series, please visit: https://www.routledge.com/Routledge-Studies-in-the-Philosophy-of-Science/book-series/POS

The Aesthetics of Scientific Experiments

Edited by
Milena Ivanova and Alice Murphy

Routledge
Taylor & Francis Group
NEW YORK AND LONDON

First published 2023
by Routledge
605 Third Avenue, New York, NY 10158

and by Routledge
4 Park Square, Milton Park, Abingdon, Oxon, OX14 4RN

Routledge is an imprint of the Taylor & Francis Group, an informa business

ISBN: 978-1-032-20388-1 (hbk)
ISBN: 978-1-032-20507-6 (pbk)
ISBN: 978-1-003-26392-0 (ebk)

DOI: 10.4324/9781003263920

Typeset in Sabon
by KnowledgeWorks Global Ltd.

Contents

Contributors

Claire Anscomb, Lecturer, De Montfort University, UK

Adrian Currie, Senior Lecturer, University of Exeter, UK

Adrian Holme, Senior Lecturer, University of the Arts London, UK

James Nguyen, Assistant Professor, Stockholm University, Sweden; Institute of Philosophy, School of Advanced Study, University of London, UK; Centre for Philosophy of Natural and Social Science, London School of Economics and Political Science, UK

Glenn Parsons, Professor, Toronto Metropolitan University, Canada

Sophie Ritson, Research Strategy Project Officer, University of Melbourne, Australia

Michael T. Stuart, Associate Professor of Philosophy, National Yang Ming Chiao Tung University, Taiwan; Research Fellow, London School of Economics, UK; Lecturer in Philosophy, University of York, UK

Alexander Wragge-Morley, Lecturer, Lancaster University, UK

Introduction

Milena Ivanova and Alice Murphy

0.1 Aesthetic Values in Science

In recent years, the relationship between aesthetics on one hand, and science on the other, has generated substantial interest in philosophy. Several collections have emerged, exploring the connections between these domains. In 2010, Roman Frigg and Matthew Hunter published an edited collection *Beyond Mimesis and Convention* that examines the notion of representation in science and art. Broadening the focus quite significantly, in 2018, Otavio Bueno, George Darby, Steven French, and Dean Rickles published the first edited collection to examine themes in the philosophy of science and aesthetics, titled *Thinking about Science, Reflecting on Art: Bringing Aesthetics and Philosophy of Science Together.* In this collection, the authors engage with the notion of representation in arts and science, the relationship between scientific models and works of fiction, the act of interpretation in art and science, and the question of whether there can be a science of aesthetics. Very recently, Milena Ivanova and Steven French's 2020 edited collection, titled *The Aesthetics of Science: Beauty, Imagination and Understanding*, offered a number of new directions in the debate specifically on topics concerning aesthetic evaluations in the scientific context, such as how beauty affects the evaluation of scientific products, the relationship between beauty and truth, but also the acquisition of understanding, and the experience of the sublime in science.

In addition to this gaining of momentum in philosophy of science, sociologists have explored how contemporary scientists experience the aesthetic dimension of their work. Very recently Jacobi et al. (2022) have identified the positive role aesthetic experiences play in scientists' well-being. In the largest empirical investigation of scientists' aesthetic attitudes to date, the study identifies a significant positive correlation between encountering beauty, awe, and wonder in science and feeling more job satisfaction and personal flourishing. While the implications from this study are yet to be investigated, it has long been acknowledged that beauty is a source of great motivation for scientists, by scientists themselves (Chandrasekhar 1987; Poincaré 2001; Hossenfelder 2018)

DOI: 10.4324/9781003263920-1

and by philosophers (McAllister 1996; Ivanova 2017; Ivanova and French 2020).

It is clear that the relationship between aesthetic values and science is now systematically studied in a number of fields and is therefore beginning to receive the attention it deserves. As a result, we have more insight into scientific methodology and the role of values within scientific practice. However, in this emerging field, a central aspect of scientific practice – the experiment – is largely missing.

0.2 The Aesthetic Dimension of Scientific Experimentation

The literature on aesthetic values in science has concentrated on their place in the evaluation of scientific theories. Crucial questions have been investigated: Are beautiful theories more likely to uncover the truth of reality? Can they be a central part of a theory's success? Or are they at best motivation for scientific pursuit? Further, could beauty be, in fact, a hindrance to epistemic goals, a bias that ought to be eliminated from scientific pursuits? This narrow perspective reflects a more general tradition in the philosophy of science of an almost exclusive focus on theorising in philosophy of science. Works such as Ian Hacking's (1983) *Representing and Intervening*, Allan Franklin's (1986) *The Neglect of Experiment*, and Deborah G. Mayo's (1996) *Error and the Growth of Scientific Knowledge* reacted to this, arguing that philosophy of science had ignored important epistemological questions emerging in experimentation and that the experiment has an autonomous life from theory.

Similarly, in the very recent literature in aesthetics of science, we are starting to see a welcomed turn to various aspects of scientific practice and their aesthetic dimension. Attention is now paid to aesthetic values in palaeontology, such as in the reconstruction of fossils and the understanding of landscapes (Turner 2019; Wiley 2021), chemical biology, including the beauty of molecules, molecular images and synthesis (Parsons 2012), anatomy (Clarke and Ambrosio 2018), and thought experiments (Murphy 2020). Crucially, in an important paper, Parsons and Rueger (2000) defend the idea that scientific experimentation can be a source of aesthetic appreciation, identifying a number of components that make an experiment aesthetically pleasing, from simple and economic design to the uncovering of beautiful natural phenomena. More recently, Ivanova (2021, 2022) has further developed an account of aesthetic value of experiments, distinguishing between immediately accessible visual beauty and the more intimate intellectual beauty unveiled in a beautiful apt design for purpose. Defending a pluralistic understanding of significance of experimental results, from experiments whose results align with expectations, to those that disrupt and challenge, she has argued that the ultimate beauty in experiments is to be found in the interplay between design and significance.

This book aims to investigate a number of questions that are yet to be systematically investigated in the literature. First, what makes an experiment beautiful? Are there some specific features that make us value an experiment aesthetically? And if we answer in the positive, are these features stable, or do they change depending on experimental tradition? Since James McAllister's seminal work *Beauty and Revolution in Science*, it has been argued that beauty judgements in science can vary, and the notion of beauty is unstable. While McAllister defended the dynamicity of beauty judgements in different theoretical traditions, should we make similar inferences when it comes to aesthetic judgements in experimental practice? It is undeniable that experimental practice has changed dramatically over the centuries. It once involved one scientist, a small room, and a few tools. Today, thousands of scientists are involved in the construction of an experiment, its location can surpass the boundaries of a country, and the information that an experiment generates cannot be accommodated by a single monograph. If the practices of conducting an experiment can change, should we also assume that the way that experiments are evaluated in terms of their aesthetic qualities can also change?

Second, what role do aesthetic factors play in the construction of an experiment and its reception by scientists as well as by the public? Can the beauty of an experiment affect how successful it is considered to be or whether it should be pursued in the first place? It is easy to assume that an experiment is beautiful only when it shows something new, when it expands our knowledge of the world, for instance, by making a discovery. But this conception seems limiting since experiments are performed with many different aims in mind: to test theories, to study phenomena, to replicate and corroborate results, to explore without a presupposed theoretical framework, and to more broadly gain understanding. Is there, then, a more fundamental goal that seems to be accomplished by a beautiful experiment?

Third, we demonstrate how a focus on beauty is too narrow when exploring aesthetics in science. We will explore a fascinating case of experiments designed and run by Artificial Intelligence (AI) using deep learning algorithms. A number of important questions arise in these new cases in science, for instance, can such experiments have aesthetic value at all? Various contributions consider the role of negative aesthetic values in science, including "ugly" experimental results, as well as how the concept of the sublime and the profound can be used to explain the significance of unusual and unsettling phenomena in science.

Fourth, through a study of experiments in the Royal Society, and in particular through a close look at Joseph Wright's painting, An Experiment on a Bird in the Air Pump (1768), we analyse the depiction of scientific experiments in artworks and turn to the history of science to study the commonalities between the conduct of experiments and artistic performances. While it seems true that today, experiments are clearly delineated from artworks, we consider whether art itself can be said to serve as an experiment. If so, does this shed new light on experimental practice? And can the connections between artworks and experiments be used to advance discussions on the cognitive value of art?

With these diverse contributions, we hope to significantly advance the emerging literature on the role of aesthetics in science. The collection offers novel insights into old questions, as well as presenting previously unstudied issues that necessitate future research, from philosophers, historians and sociologists of science, as well as scientists themselves. This interdisciplinary engagement will further our understanding of the operations of science and the values that shape and guide it.

0.3 Summary of Contributions

In the chapter, "The Aesthetic Value of Scientific Experiments", Milena Ivanova asks which aspects of experimental practice we find aesthetically pleasing and offers a classification of contexts in which scientists report to find experiments aesthetically valuable. Starting with more visual features, she argues that the subject under study in the experimental setting can be beautiful, but so can the instruments, tools, and technologies used in the experiment. Aesthetic value is also found in the design of experiments and their results. Here, Ivanova argues that beautiful experimental design can mean simple, elegant, and economical design, but in the context of modern scientific experiments, it is more appropriate to see beauty in the aptness and suitability of the experiment for its purpose. Results can be aesthetically appraised as well, importantly not only when they deliver on expectation, by confirming a theory under testing or detecting a particle or phenomenon that has been theoretically derived, but also when the experiment delivers experimental results that productively challenge established assumptions and promote further investigation. Finally, aesthetic value is also to be found in the very performance of experiments, being particularly feature of early scientific traditions, and in the creative process that leads to coming up with experimental design, conceptualising the tools needed and producing the ultimate set-up where aesthetic judgement is integral to the process.

Ivanova considers the implications of this framework for the current discussion on the nature of aesthetic judgement in science and whether the account is compatible with a dynamic understanding of beauty or a stable one. While each of the six categories is rather flexible and accommodates diverse experimental traditions, the relationship between design and significance remains as a source of deep aesthetic value that can be instantiated in different ways.

The above discussion leads to the next contribution, exploring the ever-changing nature of the concept of beauty in science and in art. In her "Beautiful Experiments in Art and Science", Claire Anscomb presents a striking asymmetry between the aesthetic evaluation of scientific experiments and those concerning "experimental" art. Whereas scientific experiments are taken to be beautiful when they elicit pleasurable, aesthetic responses due to features such as their elegant and economical design, the avant-garde art of the 20th century was celebrated for its very rejection of beauty, which had historically dominated western art. The work of artists such as Duchamp and Dada resisted traditional methods of analysis; they could not be appreciated (at least not properly) for their perceptible properties. Rather, their aesthetic value lies in their meaning. While Anscomb tracks a change in both the aesthetic appreciation of experiments and in art, in terms of a shift towards how an experiment or an artwork relates to some theoretical or conceptual claim, she argues that 'the scientific experiments are highly praised as beautiful, while the artistic experiments are not' (p. 43). This sets up the central question of the chapter: is there some conception of beauty that can help explain this difference?

Anscomb first considers a Kantian notion of beauty as "disinterested, sensuous pleasure", but ultimately argues that it is ill-suited to these practices. A Kantian view of the beauty of experiments would imply that the result of an experiment is the source of aesthetic value, which overlooks the centrality of experimental design in our appreciation of experiments. In the case of experimental art, a Kantian perspective might be able to capture how conceptual works prompt imaginative engagement but it implies that this activity would lead to pleasure, which is at odds with many experiences with conceptual art. In light of this, Anscomb turns her attention to an alternative conception: functional beauty. Functional beauty is experienced when we appreciate how an artist has achieved a goal with the materials of choice, being revealed ultimately through a consideration of the choice and message rather than in the visual features of the artwork. The concept of functional beauty has the ability to accommodate for avant-garde art and conceptual art because the beauty is not necessarily to be found in the aesthetic features of the object, but rather in what the artist wants to convey and how they do so. Duchamp's Fountain is not beautiful to look at, but the idea behind this ordinary object taking space in an art gallery causes a significant reaction in the audience; the work is interrogative and challenges us in uncomfortable ways. The idea behind this artwork can thus be seen as functionally beautiful. Just like the previous contribution by Ivanova, the relationship between *what* an artwork or an experiment does and *how* it does so is what is considered at the heart of deep aesthetic appreciation.

In his contribution to the volume, Alexander Wragge-Morley presents a detailed insight into Robert Boyle's (1627–1691) attitudes towards the epistemic value of experimentation over deductive reasoning from first principles. While philosophical interest in aesthetics, understood in terms of the affective states that arise from our sensory experiences of objects in the world, did not take off until a century later, the shift to experimentation raised questions regarding the senses (and the pleasures and pains that they evoke) as a legitimate foundation for science. Rather than steering clear of evocative descriptions to try and dissuade such scepticism, Boyle frequently referred to his own pleasurable experiences of chemical experimentation in his writings. Wragge-Morley sets out to show that for Boyle, there was an intimate connection between the pleasures produced by a literary style and the generation of scientific knowledge. This interpretation is encapsulated in two arguments.

First, Boyle described his experiments in such a way as to not only aid the reader's formation of vivid, mental images (which in turn compelled assent) but to also provide a rejection of the Aristotelian explanation for why substances are the way that they are, that is, hylomorphism. Boyle provided an alternative theory, the corpuscular hypothesis, which held that the sources of a substance's properties are the compositions of, and interactions between, small imperceptible particles of matter. While corpuscles themselves could not be perceived, Boyle argued that access to corpuscles was available via an embodied interaction with ordinary objects of experience. Wragge-Morley highlights how such a perspective has relevance for contemporary discussions of aesthetics in science, given the emphasis on the role of the body, and its likes and dislikes, in cognition. In addition, Boyle made analogies between mechanisms familiar from sensory perception to behaviour at the imperceptible corpuscle level in order to show the explanatory power of his proposed mechanical model of physical causation.

Second, Boyle argued that the pleasurable, sensory experiences brought about by his experimental results demonstrated their significance. Here, Wragge-Morley highlights place in Boyle's writings where he is more explicit in his connection to the aesthetic realm. Boyle was concerned that the artificial nature of the experiments he described might lead people to consider the substances he dealt with as irrelevant to how the natural world worked. Boyle resisted this line of thought. He stated that beauty allowed an experimentalist to separate natural phenomena from artifacts. But the substances he produced were just as beautiful as their counterparts outside the laboratory. Beauty is therefore crucial for Boyle's argument that his experiments provided evidence for the corpuscular theory: the means through which he produced crystals resulted in objects indistinguishable, and as beautiful as, those occurring in nature.

Like Anscomb, Adrian Holme's contribution to the volume focuses on the experiment in art. In his chapter, "Science as Revelation", he presents a detailed analysis of a famous depiction of an experiment: Joseph Wright of Derby's painting *The Alchymist in Search of the Philosopher's Stone, discovers Phosphorous, and prays for the Successful Conclusion of his Operation, as was the Custom among the Ancient Chymical Astrologers* (first exhibited in 1771). Holme explores why, given that alchemy had long been debunked and the field of chemistry well-established, did Wright choose such a subject? From the title, we know that the painting shows a discovery of a secular, scientific phenomenon. So why did Wright depict it in such a religious context? Furthermore, what can the painting tell us about our relation to science, both then and now?

Holme begins by offering a background to Wright's "Candlelight paintings". The way in which light shapes space within darkness is related to a Lockean understanding of visual perception, and Holme also emphasises how the use of light and dark produces experiences of the sublime, associated with darkness, obscurity, terror, and feelings of awe. The chapter then turns to Wright's other "scientific" works, namely, *the Lecture...* (1766) and *the Experiment...* (1767–1768) bringing to the fore multiple interpretations of these paintings and noting the centrality of the experimental apparatus, both for the viewer and for those in the scene. Most notably, Holme offers a discussion of the use of religious iconography, and an, alchymical reading of *the Experiment*, in which, for example, the white bird in the glass case also represents the process of sublimation.

With this in place, Holme returns to *The Alchymist*. Unlike the other two works, the painting depicts not a theatrical demonstration in front of an audience, but rather, a particular moment in time, when a genuine, unexpected discovery occurred in the laboratory. The religious and alchymical elements of the painting are explicitly present in the painting, rather than just acting to add drama to a scientific demonstration. The painting mixes a contemporary laboratory with a gothic setting, along with alchymical and religious iconography. A perplexing feature of the work is how the Alchymist's wonder at what he has discovered, and his understanding of it as a miraculous event, is portrayed in a sympathetic light. While alchymists were often depicted as delusional, even as charlatans, Wright's alchymist is shown to be a serious scholar who holds authority. While the alchymist clearly witnesses his discovery as a divine revelation, we, the viewer, know that it is a scientific discovery. Yet we can have a similar experience of awe when we view the work. *The Alchymist...* affords us the experience of scientific discovery as revelation.

In "Towards a new aesthetics of science: aesthetic cultures and the processes and objects of regard", Adrian Currie offers a lens through which to analyse current perspectives in aesthetics of science, and motivates an alternative viewpoint. While Currie does not focus solely on experimentation, his account (as with the other chapters in this collection) offers a departure from the aesthetics of scientific theories towards other features of scientific practice. In particular, he presents a detailed analysis of the aesthetic in the preparation of fossils. To clarify what is meant by the aesthetic in science, Currie identifies a trend towards the "low-falutin" sense of the term, that is, how 'valenced, affective experience – and the judgements that arise from those qualities of experience – connect with human activity' (p. 116). As with Wragge-Morley's paper, Currie flags up the important role of embodiment in aesthetic experiences. Such a conception is contrasted with the "high-falutin" Kantian view of disinterested pleasure.

Currie separates a selection of current views on aesthetics in science with reference to two key areas of disagreement: (1) the relation between the aesthetic and the epistemic and whether one is "in service" of the other, and (2) whether there is stability of the aesthetic, whether this is across the sciences, within particular disciplines, or at different points in history. Currie discusses each of these points in light of the concept of an "aesthetic culture", the "low-falutin" aesthetic sensibilities – the aesthetic skills, affect, norms, judgements, etc. – that form part of a scientific cultural identity. For example, on James McAllister's view, scientists' aesthetic inclinations have been developed in light of the empirical success of past theories. Thus, the aesthetic is in service to the epistemic. Whereas for Milena Ivanova, Currie argues, it's the other way round: scientists' aesthetic preferences have shaped how theories are constructed. Further, for McAllister, aesthetic cultures remain stable in normal science, but there are periods of revolution in which we see change. Whereas for Ivanova, given that aesthetic cultures are aesthetic-led, and aesthetic preferences are "deeply engrained in human thinking", we should not expect radical discrepancies. That is to say, aesthetic cultures are largely stable.

In the final section, Currie turns his attention to a methodological issue: when we discuss the relation between the aesthetic and the epistemic, and when we consider the stability or malleability of aesthetics in science, what aspect of an aesthetic community should we be focusing on? Currie argues that we ought not to focus on aesthetic concepts (beauty, elegance, etc.) nor on phenomenology (the experiences of scientists) but rather, the objects and processes of aesthetic sensibility. This comprises of both the elements of scientific practice and the objects that the practice is directed towards. It is this, Currie argues, that is central to membership in an aesthetic culture, and which impacts the production of scientific knowledge.

Murphy's chapter is centred around the question: what makes a scientific experiment "profound"? Scientists not only use classical aesthetic concepts in their evaluations of theories, experiments, and representations but also comment on the "depth" and "profundity" of experiments and their results. Profundity, Murphy argues, can be understood as a hybrid aesthetic-cognitive quality. As with Anscomb's and Nguyen's chapter (discussed below), we can thus see a connection between the aesthetics of science and a central debate in philosophy of art: the relation between the aesthetic and cognitive value of artworks.

Murphy draws on concepts of profundity, as developed in discussions of literary fiction and music in the philosophy of art, in order to cast light on the ways in which experiments and thought experiments, such as the double-slit with single electrons, Schrödinger's Cat, and Ampère's experiments on electromagnetism, are profound. These are linked to developments in philosophy of experiment, concerning the relationship between theory and experimental practice, developed by the likes of Elgin (1993), Hacking (1983) and Steinle (2016). On the first theory of profundity, profound experiments are designed to present their profound content in a way that brings about a deeper understanding of it, allowing the viewer to "see" something that would otherwise be difficult to access or comprehend. On the second theory of profundity, profound experiments are those that bring about results that cannot be straightforwardly interpreted. Such "deep" experiments, on this second view, can even necessitate the development of new theoretical and conceptual work. Murphy does not adopt one view over the other, arguing that both capture different senses in which experiments can be considered profound.

Murphy ends by demonstrating how broadening the class of aesthetic values in science to include features such as the profound highlights the potential limitations of a dominant view of aesthetic experience, namely, the fluency processing account. Such a view has been fruitfully applied to aesthetics of science and mathematics, given that it brings together the epistemic and the aesthetic via associating aesthetic experiences with feelings of "fluency", and consequently, helps to explain the (apparent) connections between beauty and truth. While the first theory of profundity can be assimilated into such accounts, profound experiments (on the second view) seem more naturally associated with feelings of disfluency rather than fluency. Yet, they also have both an aesthetic and an epistemic/cognitive dimension. This opens the door to thinking about a plurality of aesthetic and epistemic emotions, and their significance in the scientific context.

This discussion of aesthetic experiences in science and their relation to fluent or disfluent cognitive processing is further advanced in the next contribution exploring the nature of prosthetic perception. Complex scientific instruments enable the observation of beautiful, awe-inspiring phenomena. From images like Cosmic Cliffs recently constructed by the most powerful telescope science has invented to date, the James Webb Space Telescope, to intricate protein structures observed under powerful microscopes, science enables us to access distant and microscopic phenomena with the aid of instrumentation, allowing us to experience a "prosthetic" perception. Inherently prosthetic perception is often described in terms of awe-inspiring, wonderous, and even sublime experiences. In the chapter "The Sublime as Epistemic Expansion", Glenn Parsons argues that we should develop a more inclusive or broader notion of the sublime that can accommodate the diverse experiences we can have in science, including such prosthetic observations. While the traditional concept of the sublime connotes the experience of some kind of cognitive limitation or experience of boundaries of our understanding of the subject we are engaging with, Parsons argues that this notion does not do justice to many experiences in science. Claiming that an experience inspires wonder fails to ensure that we are talking about an aesthetic experience since wonder can be purely an epistemic experience.

While sublime seems more apt in describing the aesthetic dimension of many experiences we have when engaging with nature or scientific objects, Parsons argues that taking these experiences to denote cognitive failure is limiting, asking for a reconceptualisation of what sublime experiences really are. Engaging with the Cosmic Cliffs, for instance, can be positive, rewarding and exciting, leading to what Parsons calls "epistemic expansion". This does not necessarily involve cognitive failure but rather can be an experience of cognitive success.

Sublime experiences occur when we engage with objects and concepts that make us experience our own smallness, limitations, and constraints. These have a negative dimension to them – engaging with the stars, for example, triggers a feeling of smallness, a fear of our own lack of understanding of the universe, due to an inability to align our concepts and knowledge with our experience of the world. We do not experience space as expanding, so these feelings have a negative dimension. But the traditional accounts of the sublime usually bring the agents' experience back to being rewarding: when engaged with these negative feelings, the agent gains a positive experience by acquiring new understanding or by being able to successfully apply a concept to enlighten their experience. As such, the sublime has both a positive and negative dimension.

Parsons' novel account of the sublime presses further an emerging trend in the aesthetics of science. While classical debates have revolved around the question of how aesthetic factors relate to epistemic success (is beauty a guide to understanding? Does beauty track the truth?), we are seeing a shift in the literature to scientific practices more generally, including how these practices are infused with the aesthetic sensibility of the practitioners. In asking how we can understand the aesthetic dimension of inherently prosthetic perception in science and offering an answer by developing a new conceptualisation of the sublime, this work brings us to a deeper exploration of the aesthetic dimensions of science.

Philosophers have shown how importing insights from aesthetics into philosophy of science can be a productive pursuit when it comes to furthering our grasp on the role of aesthetic features like beauty, simplicity, and the sublime in science. Yet, in "Aesthetic Cognitivism, Experimental Shielding, and Explanatory Reasoning", James Nguyen argues that we ought to take the discussion in the other direction as well: reflecting on philosophical work on scientific experiments and thought experiments can illuminate the ways in which artworks have epistemic value. In particular, Nguyen argues, such a comparison can clarify how works of literary fiction provide propositional knowledge about the actual world.

Nguyen's argument is developed via two analogies between (thought) experiments on the one hand and works of literary fiction on the other. Firstly, experiments involving isolating certain features of interest and seeing how they interact. Through conducting an experiment, scientists gain knowledge about the world beyond this set-up alone. Similarly,

Nguyen argues, works of literary fiction are "shielded systems" in which features of interest can be isolated from the complications of the actual world. Nguyen presents the case of Austen's *Pride and Prejudice*. The novel explores the interaction between class, the social status of marriage, and the difficulties of understanding the actions of others, and it displays these features of interest in isolation from various factors that interact with them in the world. And, as with experiments and thought experiments, we gain knowledge of those features that generalise beyond the work of literature itself. That is to say, we learn something about how they operate in the actual world.

Secondly, experiments and models are used to explore whether an explanation that holds in one system also holds in others. If a hypothesis is robust, we have reason to accept it over competing hypotheses that are not. Works of literary fiction can also be used to choose between competing hypotheses. One of Nguyen's examples is the competing explanations concerning the power dynamics in romantic love. Based on our own experiences, we may not be able to discriminate between a hypothesis that focuses on gender, one that focuses on class, and one that is intersectional. Nguyen argues that Rooney's *Normal People* acts as a test for these competing explanations. The novel presents a state of affairs in which we see the power dynamic in a relationship between a middle-class woman and a working-class man shift when they go from one social context, that of high school, to another, that of University in Dublin. Such a shift would not be predicted by a singular gender or class-based hypothesis alone, thus the novel confirms the intersectional approach. Nguyen presents some possible objections to his view that stem from issues concerning the cognitive gains from artworks on the one hand, and their aesthetic appreciation on the other. He emphasises that the two ways in which we can learn from literary fictions are intimately tied to our appreciation of them as artworks. The chapter ends with some reflections on the shared cognitive aims of art and science, and the possibility of a unified epistemological framework across the two practices.

While most of the chapters in the collection explicitly examine what makes an experiment aesthetically valuable and beautiful, in her contribution "Something from Nothing" Sophie Ritson explores whether ugliness can play a positive role in science. The context of this question is the current failure to detect supersymmetric particles. Despite systematic attempts, there is no evidence so far that such particles exist, and the community in high-energy physics is having to reconsider the significance and implications of these null results. Ritson asks what is the epistemic and aesthetic dimension of such results? Can null results be aesthetically valued? If so, how? And how is the aesthetic experience associated with these results to be understood in relation to what they achieve?

Ritson discusses the productively disruptive quality of current results attempting to measure the self-coupling of the Higgs boson in the ATLAS and CMS experiments. The focus on measurements of the self coupling of the Higgs boson, Ritson argues, illustrates how ugliness can play a productive, even transformative, role in science by puting into question theoretical assumptions, and leaving the community to reconsider their commitments, opening new possibilities. Ritson thus sees such results as broadening the spectrum of theoretical possibility, encouraging dissent and proliferation in what can be seen to be a theoretically myopic field. While such results can thus be seen to play a positive epistemic role, insofar as they are probing new investigation and the conceptualisation of previously unconvinced alternatives, Ritson sees such results are having negative aesthetic value, as being ugly.

Ritson then defends the idea that in some situations, such as the current case in high energy physics, ugliness can have positive epistemic value. While negative results not aligning with theoretical expectations can open possibilities for new explorations (hence, having epistemic value), they are ugly. This is because they do not fit our commitments and aesthetic canons, and are thus not valued in the same way as results aligning with expectations. Contrary to Ivanova, who takes null results produced by the Michelson-Morley experiment to still qualify as aesthetically pleasing because of their significance and relation to well-set-up experimental design, Ritson identifies such results with negative aesthetic value. We find that this discussion opens some questions for further discussion: can ugly results become beautiful in the future? We saw that in the first chapter Ivanova argues that the Michelson-Morley experiment is one of the most beautiful experiments, and the non-detection of the ether was an epistemically significant result with its own aesthetic dimension. Ritson, on the other hand, argues that results are ugly. We could ask whether such ugly results could become beautiful if, in the future, they end up supporting a new theory as was the case with the null results of detecting the ether. While originally the null results of the Michelson-Morley experiment were disruptive, later they become important confirmation for Einstein's special theory of relativity. So how do we aesthetically value these results in their different roles, and does this affect the value of the experiment itself? This chapter opens a number of interesting questions that deserve further exploration, as well as subtle disagreements regarding the nature of the aesthetic experience and their cognitive value identified by other contributors in the book.

Finally, in "The future won't be pretty: The nature and value of ugly, AI-designed experiments", Mike Stuart asks whether experiments designed in the future will have aesthetic value and if so, how this will affect the production of knowledge. By exploring recent examples of experimental designs that are a product not of human design but of a machine learning algorithm, Stuart asks whether such experiments can be valued aesthetically and how this impacts their epistemic value.

Stuart argues that the value of the experimental design produced by the machine lies in how it overcomes human limitations. If we see aesthetic values as constraints on our reasoning that severely limit the kind of possibilities scientists can conceive of, then overcoming those can only have epistemic benefits. But this epistemic benefit comes at the cost of the aesthetic, rather than going hand in hand with it. Stuart argues that the more we benefit from automated experimentation, where AI will produce more and more apt experimental designs, the less beauty we will be seeing in those designs. Contrary to the previous chapter by Ritson, which focused uniquely on the (lack of) aesthetic value of experimental results, Stuart is focused on the negative aesthetic value of the design. While Ritson sees the ugly results as delivering the epistemic goods, in the experiments Stuart considers, it is also via the ugly design that we benefit epistemically. The good news, Stuart argues, is that scientists will spend less time designing and running experiments and be free to do more fun things with their time, such as conceptual work. But the bad news is that the future in the lab might not be as pretty.

This conclusion raises further questions regarding the goals of science and the role of agency in scientific practice. The primary goal of science is often seen to be the delivery of epistemic goods, and in this sense, AI-generated experiments certainly deliver knowledge about the subject matter and lead to discoveries. But we might worry whether they deliver other goods, such as understanding, puzzle-solving tools, or even the joy usually associated with the process of discovery. In the first contribution in the volume, Ivanova discusses recent results in sociology that have identified scientists relating the aesthetic experiences in the lab to a sense of flourishing, connected to their sense of self and well-being. Will the automation of activities usually construed as creative and involving human agency have negative effects on scientists? This tension deserves further exploration and brings us back to our initial problematisation: does science only have epistemic goals, or are these goals intimately linked to our aesthetic enrichment?

Through a number of different disciplinary perspectives – the history of art and science, sociology, philosophy of science, and philosophy of art – our collection has entertained the many ways in which experiments can be appreciated aesthetically. We have shown how historical figures viewed the aesthetic in their work, and how contemporary scientists think about the aesthetic dimensions of their practice. The latter includes the changes when we move to big experiments and towards automation. The collection has identified the many ways in which experiments are praised aesthetically, and shown how the source of aesthetic appreciation varies across different experiments in science and art. Furthermore, it has demonstrated the diversity of aesthetic responses experiments can give rise to and how this connects with their cognitive dimensions. While we hope to have offered many fruitful answers to previously unexplored questions, we feel that this is only the beginning of an investigation that we anticipate will be continued beyond this volume.

Bibliography

Bueno, O. et al. (2017) *Thinking about Science, Reflecting on Art: Bringing Aesthetics and Philosophy of Science Together*. London: Routledge.

Chandrasekhar, S. (1987) *Truth and Beauty: Aesthetics and Motivation in Science*. Chicago, IL: The University of Chicago Press.

Clarke, B. and Ambrosio, C. (2018) 'The nervous system and the anatomy of expression: Sir Charles Bell's anatomical watercolors', *Progress in Brain Research*, 243, 109–138. https://doi.org/10.1016/bs.pbr.2018.10.011.

Elgin, C. Z. (1993) 'Understanding: Art and Science', *Synthese*, 95(1), 13–28.

Franklin, A. (1986) *The Neglect of Experiment*. Cambridge: Cambridge University Press.

Frigg, R. and Hunter, M. (2010) *Beyond Mimesis and Convention: Representation in Art and Science*. Boston Studies in Philosophy of Science. Dordrecht: Springer.

Hacking, I. (1983) *Representing and Intervening: Introductory Topics in the Philosophy of Natural Science*. Cambridge: Cambridge University Press.

Hossenfelder, S. (2018) *Lost in Math: How Beauty Leads Physics Astray*. New York: Basic Books.

Ivanova, M. (2017) 'Aesthetic Values in Science', *Philosophy Compass*, 12(10), e12433.

Ivanova, M. (2022) 'What Is a Beautiful Experiment?', *Erkenntnis*. https://doi.org/10.1007/s10670-021-00509-3.

Ivanova, M. (2021) 'The Aesthetics of Experiments', *Philosophy Compass*, 16(3). https://doi.org/10.1111/phc3.12730.

Ivanova, M. and French, S. (2020) *The Aesthetics of Science: Beauty, Imagination, Understanding*. New York: Routledge.

Jacobi, C. et al. (2022) 'Aesthetic Experiences and Flourishing in Science: A Four-country Study', *Frontiers in Psychology*. https://doi.org/10.3389/fpsyg.2022.923940.

Mayo, D. G. (1996) *Error and the Growth of Experimental Knowledge*. Chicago, IL: University of Chicago Press.

McAllister, J. (1996) *Beauty and Revolution in Science*. Ithaca, NY: Cornell University Press.

Murphy, A. (2020) 'The Aesthetic and Literary Qualities of Scientific Thought Experiments', in M. Ivanova and S. French (eds.) *The Aesthetics of Science: Beauty, Imagination, Understanding*. New York: Routledge, pp. 146–167.

Parsons, G. (2012) 'The Aesthetics of Chemical Biology', *Current Opinion in Chemical Biology*, 16, 576–580.

Parsons, G. G. and Rueger, A. (2000) "The Epistemic Significance of Appreciating Experiments Aesthetically", *British Journal of Aesthetics*, 40(4), 407–423.

Poincaré, H. (2001) *The Value of Science: Essential Writings of Henri Poincaré* (ed. S. Gould). New York: Modern Library.

Steinle, F. (2016) *Exploratory Experiments: Ampère, Faraday, and the Origins of Electrodynamics*. Pittsburg, PA: University of Pittsburg Press.

Turner, D. (2019) *Paleoaesthetics and the Practice of Paleontology*. Cambridge: Cambridge University Press.

Wiley, C. (2021) *Preparing Dinosaurs: The Work Behind the Scenes*. Cambridge: MIT Press

1 The Aesthetic Value of Scientific Experiments

Milena Ivanova

1.1 Introduction

There are many parallels between science and art. In both areas, we can create products of great aesthetic value, engage in activities that are deeply aesthetically pleasing, and produce outcomes, whether artworks, theories, models, or experiments, using imagination and creativity. While there are clear analogies between art and science, there are also important differences: science is in the business of delivering epistemic goods, where the primary goal is to construct theories, experiments, and models that help us understand and manipulate the world around us. It is not surprising that effort in the last decades in philosophy of science has been paid to understand what the aims of science are and how they are achieved. This focus has led to the development of different views on the scientific method and the relationship between theory and evidence, the nature of representation, etc. However, in these traditional pursuits, understanding the way values affect these activities has been largely ignored. The "value-free ideal" in science emerged as a result of uncovering underlying values that were deemed unhelpful or detrimental to scientific goals, urging us to defend the idea that we should aim to have value-free science. Feminist philosophers responded to this picture by arguing that science is not, and even stronger, should not be value-free, and that our values need to necessarily influence what kind of science we want.[1] The debate on values in science in the recent decades then was framed very much around the moral and political values in science, pioneered by Heather Douglas (2009), but in that debate, aesthetic values are sadly missing.

In parallel with these discussions on the value-free ideal, some philosophers of science have been asking how aesthetic values are involved in science, and whether they should be eliminated from the scientific process or in fact are constitutive of it. An idea that we often find in science, going back to antiquity, is that beauty can be a guide to the truth (Chandrasekhar 1987). Other philosophers have argued that beauty can not only motivate scientists but also facilitate their understanding of the subject matter (Breitenbach 2013; Elgin 2020; Ivanova 2020). Others as

DOI: 10.4324/9781003263920-2

well have asked whether it is rational of scientists to use aesthetic judgement, given that aesthetic concepts can be seen as subjective and subject to change. In his seminal work *Beauty and Revolution in Science*, James McAllister (1996) argues that beauty can be an unstable concept, and just like the history of art shows us clearly that what is regarded as beautiful in different artistic traditions can change, such shifts commonly occur in science too. And yet, McAllister provides a sophisticated account to show that despite their changing nature, it is rational to use aesthetic judgements in science. For McAllister, scientists form aesthetic canons when working with successful theories and models, their trust in beauty, then, is based on the past success of such theories and models. By linking beauty to empirical success, and offering such a defence on the back of the past successes of beauty, he offers a way to justify the rationality of trusting beauty and aesthetic values more generally. More recently, Derek Turner (2019) and Adrian Currie (2020) have further emphasised the idea that scientists become "attuned" to certain aesthetic values through their practices and the products they encounter and work with more generally.

In the above discussion, we see what we can call "big picture" questions being addressed, namely, can aesthetic judgements helps us in the pursuit of epistemic goals or do they undermine them? Can beauty indicate truth, empirical success, understanding or does it lead us astray? While very important, in the more recent literature, we are seeing a welcomed expansion of the kinds of questions being addressed, with more work now focusing on exposing how aesthetic values shape specific scientific practices, such as: how do aesthetic values influence the reconstruction of fossils (Wylie 2015, 2021; Turner 2019; Currie 2020)? How do aesthetic values affect studies in chemistry and the discovery of new molecules and substances (Ball 2005, 2021; Parsons 2012)? Can experiments be aesthetically appreciated and what makes an experiment beautiful (Parsons and Reuger 2000; Ivanova 2021a, 2022a)? These works uncover the aesthetic features of various aspects of scientific practice. In doing so, they give us more food for thought when it comes to addressing big-picture questions but also result in further insight into the diverse ways in which aesthetic judgements feature in science.

In this chapter, I explore the levels and contexts in which we find aesthetic judgements in experimental practice. My goal is to explore how scientific experiments are appreciated aesthetically throughout different traditions, whether change in scientific experimentation as a practice, or in the tools and techniques that are used, brings with itself a new way of aesthetically appreciating the experiment, or whether we can say some aspect of beauty remains stable throughout such different traditions. As such, this chapter provides a classification of different ways in which experiments are appreciated aesthetically and illustrates this categorisation with a number of experiments from the history of science. While these experiments belong to different traditions, my goal is to defend a unified way of appreciating those experiments aesthetically, by focusing on the intimate beauty uncovered when we appreciate experimental design well suited for purpose. My analysis also illustrates that different aesthetic aspects of the experiment serve different purposes and lead to a more optimistic take on the stability of beauty in science.

I start this chapter with three case studies. In Section 1.2, I explore the beauty of Léon Foucault's pendulum; in Section 1.3, I discuss "the most beautiful experiment in biology": the Messelson-Stahl experiment; and in Section 1.4, the beauty of the Michelson-Morley experiment. In each of these sections, I identify the different ways in which the experiments are aesthetically valuable. Following these cases, in Section 1.5, I reflect on the different roles beauty is taken to play in science before I provide my classification of aesthetic values in scientific experimentation in Section 1.6. Section 1.7 reflects on how my account fits the pre-existing debates on the dynamic nature of aesthetic values and what we can learn about this debate by paying attention to the experiment.

1.2 Léon Foucault's Pendulum

Let us start our exploration of how experiments are appreciated for their aesthetic value by recalling Léon Foucault's well-known pendulum experiment. The experiment was performed inside the beautiful Pantheon in Paris in 1851. The goal of this experiment is to demonstrate a long-contested fact: that earth rotates around its axis. To achieve this goal, Léon Foucault hung a heavy brass sphere from a long rope fixed to the inside of the dome of the Pantheon. Underneath the pendulum, he placed sand and then swung the pendulum into motion. The pendulum moved slowly back and forth tracing lines in the sand beneath it. After a long period of time, the experiment revealed that the traces in the sand were not aligned, demonstrating the earth's rotation beneath the pendulum.

Why does this experiment qualify as an example of a beautiful experiment in science? To start with, there is certainly something rather awe-inspiring and beautiful in the visual features of the experiment. Set-up in one of the most beautiful buildings in Paris and resembling a kinetic sculpture with a mesmerising back-and-forth motion, this is a very visually pleasing experiment. This visual beauty strikes the senses with immediateness. Foucault's pendulum was also displayed to the public, who would visit the Pantheon to admire the experiment. The experiment, like many others before it, was a public spectacle with an immediately accessible to the senses beauty.

But the visual beauty of this experiment is certainly only one of the ways in which we can appreciate the beauty of this experiment. I think what makes Foucault's pendulum an example of a beautiful experiment in science goes way beyond its visually pleasing features. The experiment accomplished a very significant goal: it showed the effects of earth's rotation, a phenomenon that had not previously been demonstrated but has been much disputed. In addition to its significant result, the experiment was beautifully designed too. The demonstration of such a significant fact was accomplished with an imaginative, economic and elegant design, using simple and economical materials for the purpose. I think it is this relationship between the design of the experiment and its significance that presents a very different kind of beauty, an intellectual beauty, which becomes unveiled once we engage more systematically with the experiment. While the visually pleasing features function as an invitation, they invite engagement and fascinate us to discover more about the experiment; we uncover a more intimate, intellectual beauty of the experiment once we appreciate what the experiment accomplishes and how the experimenter set-up the experiment to achieve the goal, their creativity, and ingenuity.

1.3 DNA Replication

In the famous paper "Molecular Structure of Nucleic Acids", published in *Nature* in 1953, James Watson and Francis Crick end their contribution with an important insight: 'It has not escaped our notice that the specific pairing we have postulated immediately suggests a possible copying mechanism for the genetic material' (1953a, 737). What they had defended in the paper was the double helix structure of DNA, alluding to a possible replication mechanism, which later in the year they fleshed out in their second (1953b) paper and labelled "semi-conservative". The idea behind this was that during replication, each strand of the parent DNA unwinds from each other and acts as a template for the synthesis of new complementary stands of daughter DNA, resulting in two newly synthesised DNA molecules each presenting one original strand. However, this was not the only proposal of how DNA might replicate.

Another contender, due to Max Delbrück (1954), became known as the dispersive replication mechanism. Delbrück hypothesised that the parent DNA chains break at intervals, with the parental segments combining with new segments to form the daughter DNA, resulting in two daughter helices from each parental helix, with each strand in the daughter helices consisting of alternating parental and daughter DNA. An alternative replication mechanism, that did not involve the DNA strands to unwind and separate during replication, was proposed by Gunther Stent (Delbrück and Stent 1956). According to this conservative replication, each of the two strands of the parent DNA molecule is completely preserved in the newly synthesised daughter DNA.

Determining which of these mechanisms is correct was the next priority of the scientific community, and in 1958 Matthew Meselson and Franklin Stahl published the results from their experiment, now often referred to as "the most beautiful experiment in biology". What they had done to determine the correct replication mechanism was to feed bacteria nutrients containing a heavy nitrogen isotope that is incorporated into the bacterial molecules through metabolising. Then they fed bacteria light nitrogen isotope, which allowed them to study the density of the bacterial DNA. They studied the genetic material through the next generations, knowing the rates at which bacteria multiply. Instead of using radioactive labelling of the DNA strands, common at the time, they decided to use density and separate the heavy from light DNA using density-gradient centrifugation. By studying the ratios of light, heavy, and hybrid DNA strands they obtained, they were able to eliminate the conservative and dispersive replication hypotheses, confirming that DNA indeed replicates semi-conservatively.[2]

This experiment is often regarded as the most beautiful experiment in biology and one of the reasons for which it is so celebrated is because it is an example of a crucial experiment in science (Ivanova 2022a). The experiment is regarded to have conclusively settled the question of DNA replication by giving clear confirmation of the semi-conservative mechanism while eliminating the two other competitor mechanisms of replication. For many, it is this ability to deliver a decisive answer to the question of how DNA replicates that makes the Meselson-Stahl experiment beautiful. James Watson himself argues that the experiment gave a simple answer to a question (Holmes 2008, 411). Historian of science Ernst Peter Fischer also supports this view, claiming that 'the Meselson-Stahl experiments speak for themselves and made all further commentary superfluous' (1991, 21). Fredric Holmes further argues that the simplicity and clarity of the result makes this experiment easily presentable to students, serving as an exemplary experiment for science education (2008, ix).

But the significance and clarity of the experimental results is only one side of the experiment's beauty. Of importance to determining the ultimate source of aesthetic value of the experiment is not only *what* the experiment did but *how* it did it. The experiment has an elegant, original, and apt design, optimally suited for the purpose it was set out to achieve. The very idea behind the experiment is considered beautiful and ingenious. The use of density as a way of labelling the genetic material was an original feature of the experiment, allowing Meselson and Stahl to study how the genetic material changed through the next generations of bacteria they studied. Ernst Peter Fischer argues that 'There is something beautiful in this idea alone' (1991, 21). The idea behind the experimental set-up is original, elegant, and optimal, showing innovative and creative thinking as well as aptness by delivering on the job it was designed to do. Just like our first example, the pendulum experiment, the ultimate beauty of the Messelson-Stahl experiment is to be found in appreciating the relationship between the design of the experiment and the lessons it delivers, decisive results that answered a really important question.

We have examined two notable examples of beautiful experiments in science that did something different: while the former demonstrated an important concept, the latter confirmed a hypothesis among three proposed contenders. While serving different goals, these experiments have something in common: they are both cases of experiments producing results aligning with theoretical expectation; they confirm or demonstrate hypotheses or theories we have already entertained. But not every significant result in science is one of theoretical alignment. In many cases, the significance of a result is to be found in its unexpected or surprising nature. In the next section, I want to discuss the beauty of the famous experiment designed by Albert Michelson and Edward Morley to detect the velocity of the earth relative to the ether to illustrate why it is a beautiful experiment despite its controversial results.

1.4 Detecting the Ether

By the end of the 19th century, the scientific community largely committed to the wave theory of light, according to which light is transmitted in a medium that became known as the luminiferous ether. The question that puzzled the scientific community was the nature and properties of this substance through which waves propagate. At the time it had been suggested that if one assumes that the velocity with which light propagates in the ether is the same in all directions, then the first-order effects that could be measured to determine the relative velocity of the earth through the ether would cancel out, but very small second-order effects could be measurable. To make such a measurement possible, a highly precise instrument was needed and a well-set up experiment, which was what Albert Michelson set out to do. The beautiful idea behind the experiment was the realisation that they could send two light beams of similar length across two different perpendiculars to each other paths, thus ensuring the two beams could be differently affected by their journey through the ether. What was surprising and unexpected for Michelson and Morley who implemented the experiments was that they did not detect significant relative effects. The experiment was systematically revised to account for potential sources of interference, from the temperature variations of the two revolving arms of the interferometer, which Michelson dressed in paper boxes to control for temperature fluctuations, to elevating the interferometer to a mountaintop or placing it under water. Despite the progressively more precise nature of the experiment, the results did not change significantly. As Gerard Holton (1969) reflects:

> Its fascination, which has been felt equally by textbook writers and research physicists, derives from its beauty and mystery.
>
> Despite the central position of the question of ether drift in late-nineteenth century physics, nobody before Michelson was able to

> imagine and construct an apparatus to measure the second-order effect of the presumed ether drift. The interferometer was a lovely thing. Invented by the twenty-eight-year-old Michelson in response to a challenge by Maxwell, it was capable of revealing an effect of the order of one part in ten billion. It is to this day one of the most precise instruments in science, and the experiment is one that carried precision to the extreme limits. Einstein himself later paid warm and sincere tribute to Michelson's experimental genius and artistic sense.
> (1969, 136)

The experiment has often been praised for its beautiful design and careful execution; with Einstein himself arguing that Michelson was like an artist in science, paying attention not only to devise a good but also a highly beautiful experiment. Holton (ibid., 136) notes that the experiment is beautiful both because of its design and execution, showing the meticulous way in which Michelson continued to refine the method of testing to reduce sources of error. But when it comes to its results, how can we understand them in light of what experiences they afford? As Holton explains, beyond the beauty of the design, the results were 'enormously puzzling' (ibid., 137). The results violated the expectations of the community, and Michelson himself struggled to accept the implications of his experiment. While Hendrik Antoon Lorentz and George Fitzgerald took the results very seriously, proposing independently a contraction hypothesis to accommodate the experimental results, many in the community were unsure what to make of the results. What the null results generated was a prolonged problematisation in the community, since they were putting into question an assumption that has been accepted as a fundamental truth: that the ether exists.

Despite not delivering on expectation, I take this to be a beautifully designed experiment that delivered highly significant results, but the aesthetic response we have to those is not analogous to the one we have then results align with theoretical expectation. Discovery in science is often associated with an awe-inspiring beautiful moment when our expectations meet experimental outcomes. But in cases where an experiment delivers a productive disruption, we can also experience aesthetic responses, but those would be more analogous to the responses we have to very disruptive revolutionary works of art that challenge our fundamental assumptions. I think there is scope for us to accommodate both the results aligning with expectation and the productively disruptive results that do not align with our theoretical frameworks as aesthetically valuable while recognising the aesthetic responses they give rise to are different.[3] As such, the Michelson-Morley experiment is an example of a highly elegant, innovative design that delivers a significant outcome. The innovative nature of the experiment is sticking; the interferometer itself, as an invention, is one of the most elegant and simple instruments ever invented in science and used today for many discoveries in physics. The experiment was well set-up to achieve its goal, were there an ether, this experiment was going to detect it. Instead, its profound results[4] fuelled productive problematisation in the community and later aided physicists to accept Einstein's special theory of relativity.

1.5 Levels of Aesthetic Appreciation in Science

With the above cases in mind, we should ask the very general question first: where do we find aesthetic value in scientific practice? In Ivanova (2017a), I distinguish between three levels of aesthetic judgements made in science: *objects*, *products*, and *process*. That the objects of scientific enquiry can exhibit aesthetic properties and evoke in us aesthetic experiences is easily motivated: nature, after all, allows us to engage with plenty of pleasing phenomena, entities and processes.[5] From the stunning rosy veincap mushroom (*Rhodotus palmatus*) we can encounter in our forest hikes, the colourful rainbow lorikeet (*Trichoglossus moluccanus*) we can see flying in the sky, to the perfectly hexagonal honeycombs in a beehive, nature is a source of much aesthetic experiences. But such aesthetic responses are also a product of engaging with nature under carefully crafted experimental settings, such as the observation of the process of crystallisation under a powerful microscope or the decomposition of light through a prism.

The products scientists create are also subject to aesthetic appreciation. Scientific theories, models, and mathematical proofs are very often praised for their beauty and elegance. In theoretical physics, much discussion is given to the beauty of Einstein's theories of special and general relativity. His principle theories are regarded by many to be the culmination of beauty in science, since the theories achieve great unifying explanatory power under a small set of scientific principles (Ivanova 2020). Galileo Galilei is often praised to have devised a highly elegant thought experiment, which utilises economic concepts and materials to showcase an important novel concept: that of acceleration (Murphy 2020). Similarly, scientific images are also subject to aesthetic judgement and appreciation. The recent images produced by NASA's James Webb Space Telescope (JWST) are an excellent example. Images such as Cosmic Cliffs inspired awe and wonder in the public with their beauty. To produce such beautiful images, scientists had to make a number of judgements of aesthetic nature when translating the data collected by the telescope, to use colours not detected by the telescope.[6]

In addition to the objects of study and products of scientific activity, the very process of developing a theory, mathematical proof, designing an experiment or preparing an image, involves creativity, imagination, and aesthetic sensibility. Such practices are often paralleled with artistic production, and the involvement of the aesthetic sensibility is as much a feature of contemporary science as its early predecessors. Scientists as far back as Leonardo da Vinci and Robert Hooke would enhance the aesthetic features of their subject matter in their depictions, generating in the viewer awe and wonder when engaging with the depicted specimen. Similarly, today as we discussed above, astronomers at NASA use colours to create the beautiful images obtained with the JWST. Aesthetic judgements are integral to many scientific fields, from astronomy to biology, and medicine to chemistry and palaeontology. But also the very process of scientific discovery is often compared to artistic production. Pierre Duhem, for instance, argues that it is impossible to follow the process of constructing a physical theory 'without feeling keenly that such a creation of the human mind is truly a work of art' (1954, 24). Similarly, Ernest Rutherford argues that 'the process of scientific discovery may be regarded as a form of art', continuing that 'a well constructed theory is in some respects undoubtedly an artistic production' (quoted in McAllister 1996, 14). In addition to seeing the creative process in science as parallel to artistic creativity, Henri Poincaré (2001) explicitly devises an account of scientific creativity at the heart of which is the aesthetic sensibility of the scientist, which acts as "the delicate sieve" that scans and evaluates which ideas that the mid produces are useful by judging first their aesthetic appeal. At the heart of this account is the idea that 'care for the beautiful is care for the useful' (Ivanova 2017b).

To take stock, there are three general contexts in which aesthetic judgement enters in science: object, product, and process. But when thinking about the scientific experiment specifically, we can further delineate more specific categories in which we find aesthetic value in the experiment. In what follows, I propose a categorisation of the levels at which we can appreciate an experiment aesthetically and then draw some conclusions about their function.

1.6 A Classification of Aesthetic Aspects of Experiments

Earlier in this chapter, we followed three celebrated experiments and explored their aesthetic dimensions. These examples help us illustrate the different ways in which we can appreciate an experiment aesthetically. Specifically, here I introduce six different categories at which we make aesthetic judgements and aesthetically appreciate the scientific experiment. After introducing these categories, I argue for the special relationship between design, significance, and ingenuity displayed in a well-designed experiment for purpose.

1.6.1 Subject of Study

The phenomena that are investigated under experimental conditions can be highly aesthetically pleasing. From cells, light, and electricity, to the beautiful structures that can be generated in the process of electrodeposition, scientists report finding a lot of visual beauty in their subject, be it in intricate shapes, colours, or patterns. Such features seem to be particularly prominent in some fields, such as chemistry (Ball 2005, 2021; Parsons 2012), as well as biology and astronomy (Ivanova 2022b). Scientists report that the beauty in the subject they study is a great source of beauty, awe, and wonder. Ball's beautiful study of the aesthetic dimension of chemical experimentation is a great illustration of how chemists value the sensual aspects of their work, being motivated not only by visual beauty but also by other sensory experiences in the lab. The aesthetic dimension of scientific studies acts not only as a motivator for their work but also contributes to scientists' work satisfaction and overall well-being. As a very recent study on scientists' aesthetic attitudes and well-being reported (Jacobi et al. 2022), regularly encountering beauty in the lab contributed to scientists' well-being and sense of flourishing.

1.6.2 Instruments

The tools used by the experimenter can have their own aesthetic value as the images in Figure 1.1 and 1.2 show us. From beautifully crafted microscopes and particle chambers to complex structures like the large hadron collider or an algorithm, the instruments and technologies utilised in the experiment can display craftsmanship and pleasing aesthetic features that can be appreciated in their own right.

The technology used in the lab can be both visually pleasing and appreciated for how it was created, giving rise to experiencing a more intellectual form of beauty. As we noted earlier, the interferometer, originally invented to detect ether drift and now used for many precise measurement experiments, is a highly beautiful instrument, comprising minimal elements while enabling the study of complex phenomena and detecting very small effects. It continues to be used to date in the construction of more and more sophisticated and precise technology, like the recent Machine Learning Assisted Interferometric Structured Illumination Microscopy construed by E.N. Ward (2022). Similarly, the recent JWST was also regarded as highly beautiful. The beauty of this powerful telescope was found both in its visual features, with its intricate hexagonal gold plates, as well as in the more intimate beauty that is revealed when we appreciate the highly complex and timely effort behind building such a precise and innovative technology.

1.6.3 Experimental Design

As we saw in the opening first sections of this chapter, experiments have been consistently praised for their elegant, economical, and simple set-up, as well as their aptness. The idea of aptness is especially important when it comes to experiments in the context of big science.

Figure 1.1 Image (©) Whipple Museum, Cambridge (Wh.0196).

Experiments in modern settings can be incredibly complex. Take, for example, the experiments at the Large Hadron Collider at CERN. The experiments are running on four large detectors (ATLAS, CMS, ALICE, and LHCb), the data is generated and analysed using machine learning algorithms, and the scientists working on the experiment (over 2500 of them) are located across many countries. This collaborative experiment could not be further from the experimental setups we discussed in the early days of experimental practice, where the experiment was performed by one individual, the equipment was simpler, and the experimental results were immediately accessible. In the complex set-up of today's experiments, scientists work in large teams, the data the experiment produces is not perceivable immediately, rather it involves complex and timely analysis and selection techniques, and the very set-up can seem complex by comparison. But I want to emphasise that even in the case of big experiments, beyond the complex set-up and technology, we

Figure 1.2 Image (©) Whipple Museum, Cambridge (Wh.1782).

can appreciate the optimality of the design fit for purpose. In this sense, the design remains beautiful, being appreciated for how it is set for the goal of the experimenters.

1.6.4 Creative Process

As already detailed in the previous section, constructing and running the experiment can be paralleled with artistic production. Experimenters, like artists, utilise their creativity, imagination, and aesthetic sensibility in diverse stages of experimental process. The scientists that come up with the design and those who carefully carry out the experiment out can be praised for their skill and capacity in conceiving the praiseworthy design or delivering important results. What is particularly fascinating about science today is that the creative process is shared not only among many individuals, who collectively explore possibilities to conceive of the most optimal experimental designs and tools, but they are also assisted by machines not only for running the experiment, analysing and interpreting its data but also in their very design, which opens a number of underexplored questions about the nature of creativity and coproduction with machines.

1.6.5 Significance

As we saw in our discussion of the Meselson-Stahl experiment, the experimental results can also be regarded as aesthetically pleasing. Here we can find a diversity of responses depending on the nature of the result. Experiments can produce results that align with theoretical expectation, confirming a theoretical prediction or discover something new. They can lead us to enriching our understanding but also to identify the limitations of our understanding. Surprising, unexplained, anomalous result can lead to productive problematisation which then leads us to ask new questions, generate new ideas, and potentially further our understanding of the subject matter. As noted earlier, a pleasing design without a significant result does not have the same aesthetic value, a particular deeper aesthetic value is found in the intricate connection between design and significance. And significance of experimental results can be found in those results that are expected, aligning with theoretical expectation as well as those that violate theoretical expectation. Surprising and anomalous results can be deeply profound, prompting us to productive investigations.

1.6.6 Performance

The experience of running the experiment can be seen as a process of deep aesthetic engagement. Whether it is because the experiment reveals beautiful phenomena or aptness of design, or the very feeling of being involved in the process of exploring and discovering, the experimenter as well as the audience engaging with the experiment, report having aesthetic experiences. Experiments in past traditions were often performed with audiences, becoming a public spectacle that fascinates and evokes diverse aesthetic experiences. In this volume, Adrian Holme explores the famous artworks by Joseph Wright of Darby that depict beautifully the public nature of the experiments performed in the early days of the Royal Society and the diverse nature of the aesthetic responses the audience had when witnessing the experiment. While scientific practice is now very different, and labs are rarely open for the public to observe the daily running of an experiment, we nevertheless still encounter the spectacle of science with different settings. For instance, NASA's recent initiatives to share the beautiful JWST images are analogous to the early experimenter opening the door to the public. While the setting is different, the aim remains the same, to evoke an aesthetic response in the public and fascinate them about science and its discoveries.

My argument so far has focused on identifying the diverse aspects in which we can find beauty and aesthetic value in the lab, be it in the very phenomena we study, the experimental instruments, the design of the experiment, its results, in its performance, or it encapsulating human ingenuity and creativity. In our earlier exploration of the three scientific experiments, we identified at least one of the above six categories of aesthetic value in them, but we also saw that beautiful experiments often present several of these aesthetic dimensions. Next, I want to argue for something much stronger than the idea that to value an experiment aesthetically is to appreciate one of the above six categories in which the experiment can be pleasing. Beyond defending the classification here, I want to emphasise the distinction between immediately accessible via the senses beauty and the intellectual beauty we experience upon appreciating the design of the experiment and its significance. This distinction is important not only because it directs our attention to the different ways of appreciating, but also because it motivates the idea that the immediately accessible to the senses beauty and the more intellectual beauty play different functions in science. Visual features can be great motivators for pursuit in science; they can also help in science communication and serve as didactical tools. But a deeper, more intimate beauty is found when one appreciates the creativity in designing a well-suited experiment that delivers significant results.

I have already emphasised in previous work (2022a) that by significance of results I do not only mean experimental results that align with expectation, by detecting a theoretically predicted particle or phenomenon, or by confirming a theory or a hypothesis, but also those that violate expectation. Whether it is an anomalous result, a null result, or a surprising result, such results can be the most productive experimental outcomes since they can point the finger at reconsidering particular assumptions, techniques, or experimental setups. As we discussed in the case of the ether, the null results invited a fruitful problematisation in the community. While eventually these results were seen as confirming evidence for special relativity, originally they were the anomalous result that questioned the ether hypothesis. As Hossenfelder (2018) argues, similar productive reassessment is happening now in high-energy physics with the non-detection of supper symmetric particles. The null results are not only challenging the theory that has entailed their existence, but also the aesthetic principles that have been constituting the development of physical theories in the last century, from the special and general theories of relativity, quantum mechanics, and the standard model. These very aesthetic principles of naturalness, symmetry, and simplicity, Hossenfelder argues, are at stake as well as physicists look for a way forward from the null results and failure to detect these theoretical particles. Thus, for me, it is not only that these two ways of appreciating focus on different features and serve different functions, the beauty that we encounter through the appreciation of the well-designed experiment that delivers a significant results seems to be a particularly special source of aesthetic value.

While my focus in this section has been to offer a classification of the ways in which we can come about to appreciate an experiment aesthetically, I have also defended the idea that different categories of aesthetic appreciation can play different roles, and that the relationship between design and significance is a particularly important source of aesthetic value. With this argument in mind, next I want to return to the question that has occupied philosophers for many decades: are the ways in which we appreciate aesthetically the objects of science stable or subject to change? Whether focusing on the scientific experiment can deliver new insight into this question is the subject of the next section.

1.7 Implications

When thinking about the role of beauty in science, many scientists and philosophers have expressed two very opposing views. On the one hand, some believe the concept of beauty in science is objective, stable, and agreed upon independently of the tradition, school of thought, or time period a scientist belongs to. Contrary to art, where it is believed aesthetic judgements can vary, in science, it is claimed, aesthetic judgements are stable. This Platonic view, often expressed by scientists such as Paul Dirac, gives assurances that using aesthetic judgements in science is not threatening to its objectivity, since these values are shared among scientists, and beauty can indeed guide the rational endorsement of scientific theories. This view, however, has been challenged. Is it plausible to assume aesthetic judgements enjoy stability across different time periods or even different schools or labs and aesthetic judgements in science are different to those we make with regard to artworks?

Looking at the history of art makes it clear that our aesthetic judgements can vary. Who would forget the history of the Eiffel tower in Paris, when first erected in 1887 a very large group of contemporary artists in France wrote to complain against the decision to build the tower and demand its immediate demolition. The reason was that the tower did not fit the existing aesthetic sense of the French, calling it a "black factory chimney", "an insult to the French fine taste" and a "monstrosity" over the beautiful Parisian skyline. But it did not take that long for the eye sore to become an icon and symbol of the beauty of modern architecture. The building now symbolises the beauty that can be achieved when architects use cast iron, a material that allowed for taller, more intricate, and original shapes to be constructed than ever before. How did this aesthetic transformation become possible?

One way to see the acceptance of a previously unacceptable artefact as beautiful is to argue that people's aesthetic attitude changed, and appreciation grew for the beauty of the new building. In *Beauty and Revolution in Science* (1996) James McAllister argues that just as in art, in science too aesthetic attitudes can change. And in fact, such attitudes are learned, they are a response to appreciating the success of an artefact through a tradition. Just like in art, where appreciating the possibilities that using cast iron enabled eventually resulted in aesthetic attitudes towards buildings like the Eiffel Tower to change, in science, McAllister argues, it is through our appreciation for successful theories that we shape our aesthetic attitudes and canons. So while we might initially think theories such as quantum mechanics or the standard model do not fit our aesthetic ideals, working with these theories and appreciating their success can change our idea of beauty, slowing shifting our expectation as to what is a beautiful theory.

Two crucial questions emerge from this dispute on the nature of aesthetic appreciation in science. First, are experiments appreciated differently throughout different traditions or is there stability in aesthetic judgements when it comes to the experiment? Second, if we accept beauty is a dynamic notion, as McAllister suggests, do revolutions in aesthetic attitudes regarding scientific theories coincide with such attitudes changing in how experiments are appreciated or are these independent? To address these questions we can get insight from Parsons and Rueger (2000), who argue that the way we appreciate aesthetically can be tied to a particular methodology. In their paper, they study experiments in the 17th century, arguing that at the forefront of aesthetic appreciation at the time is not the ingenuity of the experimenters or the design but rather displaying nature's beauty through the experiment. But they see a shift in aesthetic appreciation in the 19th century, where at the forefront of appreciation comes the design of the experiment, its economy, simplicity, optimality, and aptness. As they argue:

> [N]ow it is only with the assistance of the confirmed or illustrated theory that an experiment is thought to give us insight into nature. Whatever beauty is displayed in an experiment, it cannot be the beauty of nature itself; the economy of an experiment reflects the economy of our own cognitive households, not the economy of nature. This is clearly different from the view of the eighteen century natural philosopher who appreciates nature itself through the frame of the experiment.
>
> (ibid., 411)

Parsons and Rueger argue that the way experiments are appreciated aesthetically is correlated with scientific methodology – when science was more inductively driven, what mattered was the beauty of the natural phenomenon, while when science became more hypothetico-deductive, this appreciation becomes more focused on what the design does to confirm or disconfirm hypotheses. What is aesthetically appreciated in the later methodological tradition is not nature's beauty as much as the display of '"aptness" in the relation of result and tools, of plan and success' (ibid., 411). The experiment becomes beautiful due to its 'optimally suited to achieve its purpose' (ibid., 411–412). This account leaves it open that appreciation of future experiments might change as well due to change in scientific methodology, thus aligning with McAllister's thesis that sees aesthetic judgements as revisional and the concept of beauty as dynamic, subject to re-evaluation. However, Parsons and Reuger help us see that the revolutions in aesthetic appreciation of experiments are not aligning with the aesthetic revolutions at the level of how theories are appreciated. For instance, whereas McAllister sees an aesthetic revolution in the shift from circular to elliptical planetary motion in the 16th century and another one with the development of quantum mechanics in the 20th century, Parsons and Reuger see a transition in aesthetic appreciation of experiments during the mid-19th century when the focus is on the comparison of precision measurements with theoretical quantitative predictions.

A last point I would like to reflect on regards the idea that methodology can reshape how we aesthetically appreciate the experiment. It is undeniable that scientific methodology has changed dramatically. We now live in period of big data science where scientific experiments produce enormous amount of information where special selection procedures are needed before analysis takes place. Also, the relationship between theory and data has changed, in the sense that in many areas data is now leading the formation of hypotheses and identification of correlations. We also live in the age of assisted science, where machine-learning algorithms are not only running experiments and analysing the data they produce, but are even designing them. The philosophical tradition that searched to answer the question "what is the scientific method?" has been replaced by methodological pluralism. Philosophers now recognise that science adopts a plurality of methods depending on the subject, the availability of data and possibility of testing. Analogously, experimental practice has changed dramatically over the centuries and perhaps at no other time has it been more different than contemporary experimental practice. But while we observe many changes in experimental practice, it is unclear whether we should be concluding from this that the way we appreciate aesthetically the experiment has changed. Perhaps in addition to looking at the historical evolution of experimental practice, we also need to see its diverse nature from today's perspective in order to appreciate the different contexts in which we make aesthetic judgements in the lab. Above I identified six categories of aesthetic appreciation of experiments and I think these categories can be taken to be rather stable. We appreciate aesthetically the highly complex AI-generated images or images produced with highly complex microscopy just like we appreciate da Vinci's depictions of the human body. We appreciate the complex hadron collider as well as the interferometer, both visually and intellectually upon discovering how and why they are constructed. Most importantly, we continue to appreciate the deep beauty that is uncovered when we appreciate the design of an experiment, the relationship between design and experimental result, and the creative aspect of imagining and producing an ultimate experimental set-up for purpose. The experiments I discussed in the first sections of this chapter help us illustrate this point, while very different they all exemplified a beautifully designed, apt experimental set-up, showing the creative thinking of their designer, and provoked prolong admiration with their significance. While working in big groups rather than alone with a relatively small equipment, scientists today as their counterparts in the past find beauty in the very performance in the experiment and the sense of involvement and achievement this experience gives rise to. Just like one could walk into the Pantheon in Paris to admire the beauty of Foucault's pendulum, today CERN occasionally opens the doors to the public so that they can appreciate the beauty of the experiment and its magnitude (I was lucky to recently visit the ATLAS experiment and ask a lot of naïve questions the scientists who run it). In this sense, even though the very performance of experiments has changed, this daily practice can involve aesthetic appreciation from the practitioners as well as viewers continues to be an important aspect of experimental practice. In this sense, I hope to have given a convincing argument for the stability of the aesthetic categories I identified in this chapter, while also identifying the different functions beauty judgements can play when we think about the experiment.

1.8 Conclusion

In this chapter, I have argued that experiments are appreciated aesthetically on many different levels, from the visually pleasing phenomena they unveil to the instruments and tools they employ, to the ingenuity and creativity they are a product of, to the experimental design, the significance of the result, to their performance. While in each of these six contexts, we can appreciate a certain aesthetic dimension of the experiment; I also drew a delineation between immediately accessible to the senses beauty, like visual beauty, and the beauty that is revealed when we appreciate the design of the experiment and what this experiment achieves. These different kinds of beauty play a different function in science. While the visual beauty of an experiment may invite us to engage with it and find out more about it, acting as a motivator and aiding science communication, we find a deeper, more intellectual, beauty once we find how the experiment is set-up, the instruments that are used, how the experimenters constructed the design, and what the experiment achieves. The categorisation I have proposed allows us to identify continuity between different experimental traditions despite their changing setups.

Acknowledgements

I am grateful to the British Society for Aesthetics for supporting this project and the fruitful interactions that led to this book. This chapter was presented at The Aesthetic Value of Scientific Experiments Conference in Cambridge, the BSPS Annual Conference, the annual conference of the British Society for Aesthetics, and EPSA21 as well as in departments in Aarhus, Cambridge, Geneva, Oxford, Reykjavik, Turin, Uppsala, and Wuppertal. I am very grateful to the audience for their helpful comments and questions. Thanks are also due to Hasok Chang and his PhD students, for discussing this project with me and their continuous encouragement and support. I am very grateful to Alice Murphy and Claire Anscomb for the insightful discussions we have had over the last years, and last but not least to Glenn Parsons, whose work has inspired so much of mine.

Notes

1. For a recent overview of the value-free ideal, see John (2021).
2. This is a very brief description of the experiment, for a more detailed analysis, see Franklin and Laymon (2020), Holmes (2008), and Weber (2009).
3. A number of recent articles have explored the role of surprise in science (Currie 2018; French and Murphy 2021) as well as the more disruptive dimension of experimental results (Ritson 2020).
4. See Murphy (this volume) for a systematic analysis of the nature of profundity in scientific results.
5. The literature on environmental aesthetics is very rich to make justice here, influential accounts have been developed by Brady (2013), Carlson (2000), Saito (2017), and Parsons (2008).
6. In my recent work (Ivanova 2022b), I argue that the visual beauty of these images is simply an invitation to engage with the more intimate, intellectual beauty we encounter upon appreciating how the images were produced, the significance of this new instrument and the enormous effort into constructing it, and the fact it makes the audience become part of the beauty of scientific discovery.

Bibliography

Ball, P. (2005) *Elegant Solutions: Ten Beautiful Experiments in Chemistry.* Cambridge, UK: Royal Society of Chemistry.

Ball, P. (2021) *The Beauty of Chemistry, Art, Wonder and Science.* Cambridge, MA: MIT Press.

Beardsley, M. C. (1982) 'Aesthetic Experience', in Donald M. Callen, M. J. Wreen and M. C. Beardsley (eds.) *The Aesthetic Point of View: Selected Essays*. Ithaca and London Cornell University Press, pp. 285–297.

Beauchemin, P. H. (2017) 'Autopsy of Measurements with the ATLAS Detector at the LHC', *Synthese*, 194, 275–312.

Brady, E. (2013) *The Sublime in Modern Philosophy: Aesthetics, Ethics, and Nature*. Cambridge: Cambridge University Press.

Breitenbach, A. (2013) Aesthetics in Science: A Kantian Proposal. *Proceedings of the Aristotelian Society, CXIII*, 83–100.

Bueno, O. et al. (2017) *Thinking about Science, Reflecting on Art: Bringing Aesthetics and Philosophy of Science Together*. London: Routledge.

Carlson, A. (2000) *Aesthetics and the Environment: The Appreciation of Nature, Art and Architecture*. London: Routledge.

Chandrasekhar, S. (1987) *Truth and Beauty: Aesthetics and Motivation in Science*. Chicago, IL: The University of Chicago Press.

Crease, R. (2002) 'The Most Beautiful Experiment', *Physics World*, May 2.

Currie, A. (2018) 'The Argument from Surprise', *Canadian Journal of Philosophy*, 48(5), 639–661.

Currie, A. (2019) 'Creativity, Conservativeness and the Social Epistemology of Science', *Studies in the History and Philosophy of Science Part A*, 76, 1–4, https://doi.org/10.1016/j.shpsa.2018.11.001

Currie, A. (2020) '*Epistemic Engagement, Aesthetic Value & Scientific Practice*', *British Journal for Philosophy of Science*, https://doi.org/10.1086/714802.

Delbrück, M. (1954) 'On the Replication of Desoxyribonucleic Acid (DNA).' *Proceedings of the National Academy of Sciences*, 40, 783–788.

Delbrück, M. and Stent, G. S. (1956) 'On the Mechanism of DNA Replication', in William D. McElroy and Bentley Glass (eds.) *McCollum-Pratt Symposium on the Chemical Basis of Heredity*. Baltimore: Johns Hopkins University Press, pp. 699–736.

Douglas, H. E. (2009) *Science, Policy, and the Value-Free Ideal*. University of Pittsburgh Press, https://doi.org/10.2307/j.ctt6wrc78

Duhem, P. (1954 [1906]) *The Aim and Structure of Physical Theory*. Princeton, NJ: Princeton University Press.

Elgin, C. Z. (2014) 'Fiction as Thought Experiment', *Perspectives on Science*, 22, 221–241.

Elgin, C. Z. (2018) *True Enough*. Cambridge, MA: The MIT Press.

Elgin, C. Z. (2020) 'Epistemic Gatekeepers', in Milena Ivanova and Steven French (eds.) *The Aesthetics of Science: Beauty, Imagination, Understanding*. New York: Routledge, pp. 21–36.

Fischer, E. P. (1991) *Beauty and the Beast: The Aesthetic Moment in Science*. New York: Springer.

Franklin, A. (1986) *The Neglect of Experiment*. Cambridge: Cambridge University Press.

Franklin, A. and Laymon, R. (2020) *Once Can Be Enough: Decisive Experiments, No Replication Required*. Heidelberg: Springer.

French, S. (2020) 'Imagination in Scientific Practice', *European Journal for Philosophy of Science*, https://link.springer.com/article/10.1007/s13194-020-00291-z

French, S. and Murphy, A. (2021) 'The Value of Surprise in Science', *Erkenntnis*, https://doi.org/10.1007/s10670-021-00410-z
Frigg, R. and Hunter, M. (2010) *Beyond Mimesis and Convention: Representation in Art and Science*. Boston Studies in Philosophy of Science. Dordrecht: Springer.
Galison, P. (1987) *How Experiment End*. Chicago, IL: Chicago University Press.
Hacking, I. (1983) *Representing and Intervening: Introductory Topics in the Philosophy of Natural Science*. Cambridge: Cambridge University Press.
Holmes, F. L. (2008) *Meselson, Stahl, and the Replication of DNA*. New Haven and London Yale University Press.
Holton, G. (1969) 'Einstein and the "Crucial" Experiment', *American Journal of Physics*, 37, 968, https://doi.org/10.1119/1.1975219
Hossenfelder, S. (2018) *Lost in Math: How Beauty Leads Physics Astray*. New York: Basic Books.
Ivanova, M. (2017a) 'Aesthetic Values in Science', *Philosophy Compass*, 12(10), https://doi.org/10.1111/phc3.12433
Ivanova, M. (2017b) 'Poincaré's Aesthetics of Science', *Synthese*, 194, 2581–2594.
Ivanova, M. (2020) *Beauty, Truth and Understanding, in the Aesthetics of Science: Beauty, Imagination and Understanding*. New York: Routledge, pp. 86–104.
Ivanova, M. (2021a) *Duhem and Holism*. Cambridge: Cambridge University Press.
Ivanova, M. (2021b) 'The Aesthetic of Experiments', *Philosophy Compass*, 16(3), https://doi.org/10.1111/phc3.12730
Ivanova, M. (2022a) 'What is a Beautiful Experiment?', *Erkenntnis*, https://doi.org/10.1007/s10670-021-00509-3
Ivanova, M. (2022b) 'The James Webb Images Were Made to Look Beautiful', *Institute of Art and Ideas Online Magazine*, https://iai.tv/articles/james-webb-images-were-made-to-look-beautiful-auid-2222
Ivanova, M. (2022c) 'Scientific Progress and Aesthetic Values', in Yefeng Shan (ed.) *New Philosophical Perspectives on Scientific Progress*. Routledge New York and London.
Ivanova, M. and French, S. (2020) *The Aesthetics of Science: Beauty, Imagination, Understanding*. New York: Routledge.
Jacobi, C. J., Varga, P. J. and Vaidyanathan, B. (2022) 'Aesthetic experiences and flourishing in science: A four-country study', *Frontiers in Psychology*, 13, https://doi.org/10.3389/fpsyg.2022.923940
John, S. (2021) *Objectivity in Science* (Elements in the Philosophy of Science). Cambridge: Cambridge University Press, https://doi.org/10.1017/9781009063647
Mättig, P. and Stöltzner, M. (2019) 'Model Choice and Crucial Tests. On the Empirical Epistemology of the Higgs Discovery', *Studies in History and Philosophy of Modern Physics*, 65, 73–96.
Mayo, D. G. (1996) *Error and the Growth of Experimental Knowledge*. Chicago, IL: University of Chicago Press.
McAllister (1996) *Beauty and Revolution in Science*. Ithaca, NY: Cornell University Press.

Meselson, M. and Stahl, F. W. (1958) 'The Replication of DNA in Escherichia Coif', *Proceedings of the National Academy of Science of the United States of America*, 44, 671–682.

Murphy, A. (2020) 'The Aesthetic and Literary Qualities of Scientific thought Experiments', in Milena Ivanova and Steven French (eds.) *The Aesthetics of Science: Beauty, Imagination, Understanding.* New York: Routledge, pp. 146–167.

Murphy, A. (2021) 'Towards a Pluralist Account of the Imagination in Science', *Philosophy of Science*, https://doi.org/10.1086/710620

Parsons, G. (2002) 'Nature Appreciation, Science, and Positive Aesthetics', *The British Journal of Aesthetics*, 42(3): 279–295, https://doi.org/10.1093/bjaesthetics/42.3.279

Parsons, G. (2008) *Aesthetics and Nature.* London: Continuum Press.

Parsons, G. (2012) 'The Aesthetics of Chemical Biology', *Current Opinion in Chemical Biology*, 16, 576–580.

Parsons, G. and Rueger, A. (2000) 'The Epistemic Significance of Appreciating Experiments Aesthetically', *British Journal of Aesthetics*, 40, 407–423.

Poincaré, H. (2001 [1902]) 'Science and hypothesis', in S. Gould (ed.), *The Value of Science: Essential Writings of Henri Poincaré.* The Modern Library. New York.

Ritson, S. (2020) 'Probing Novelty at the LHC: Heuristic Appraisal of Disruptive Experimentation', *Studies in History and Philosophy of Science Part B: Studies in History and Philosophy of Modern Physics*, 69, 1–11.

Ritson, S. (2021) 'Creativity and Modelling the Measurement Process of the Higgs Self-Coupling at the LHC and HL-LHC', *Synthese*, https://doi.org/10.1007/s11229-021-03317-y

Saito, Y. (2017) *Aesthetics of the Familiar: Everyday Life and World-Making.* Oxford: Oxford University Press.

Todd, C. S. (2008) 'Unmasking the Truth Beneath the Beauty. Why the Supposed Aesthetic Judgments Made in Science May Not Be Aesthetic at All', *International Studies in the Philosophy of Science*, 22, 61–79.

Turner, D. (2019) *Paleoaesthetics and the Practice of Paleontology.* Cambridge: Cambridge University Press.

Ward, E. N. (2022) MAI-SIM: Interferometric Multicolor Structured Illumination Microscopy for Everybody, https://arxiv.org/pdf/2206.00532.pdf

Weber, M. (2009) 'The Crux of Crucial Experiments: Duhem's Problems and Inference to the Best Explanation', *British Journal for the Philosophy of Science*, 60, 19–49.

Wragge-Morley, A. (2020) *Aesthetic Science: Representing Nature in the Royal Society of London, 1650–1720.* Chicago: Chicago University Press.

Wylie, C. (2015) 'The Artist's Piece Is Already in the stone': Constructing Creativity in Paleontology Laboratories', *Social Studies of Science*, 45(1), 31–55.

Wylie, C. (2021) *Preparing Dinosaurs: The Work Behind the Scenes.* MIT Press.

2 Beautiful Experiments in Art and Science

Claire Anscomb

Although there is a lot of disagreement in the literature on beauty, perhaps one thing that is widely agreed upon is that beauty evokes an experience of pleasure (Hickey 1998, 16; McMahon 2005). This pleasure has increasingly come to be embraced as an important dimension of scientific practice. Scientific experiments that are judged as beautiful tend to exhibit properties like simplicity and economy. Curiously, the experimental art of modern avant-garde artists is also frequently praised for these properties yet is not reported as producing a similar sense of pleasure. Why is this? In what follows, I will examine which conceptions of beauty can accommodate the sense of pleasure experienced in relation to practices of scientific experimentation, and the relative lack of this in experimental art. In doing so, I will look to a long-neglected conception of beauty, functional beauty (FB), and argue that this can account for the differing experiences reported by appreciators of artistic and scientific experiments.

2.1 The Aesthetics of Scientific Experiments

The construction and evaluation of scientific experiments is, naturally, subject to epistemic concerns. In addition to epistemic considerations, as a number of scientists and theorists have proposed, there can also be a strong aesthetic component in these practices. However, what is considered aesthetically valuable has varied over time – arguably in line with what is considered to be epistemically valuable.

In the 18th century, as Alexander Rueger has outlined, those in the realm of experimental natural philosophy, such as Joseph Priestly and G. C. Lichtenberg remarked on the beauty of experiments, and the pleasure of this experience (1997, 305). Importantly though, the experimental natural philosophy methodological framework viewed the experiment as being 'like a frame for the display of a picture – nature's operations; and although the frame itself can be beautiful (as an artefact), it is not the focus of the spectator's attention' (Parsons and Rueger 2000, 409) – rather, this was the beauty of nature. Indeed, with this framework, theory was

DOI: 10.4324/9781003263920-3

not to be imposed upon nature, which would instead reveal itself. This approach shared affinities with aesthetic theories that were developed around a similar time, which focused on notions of disinterestedness.

Broadly speaking, on the disinterested view as promoted by theorists such as Hutcheson, the experience of beauty is bracketed from practical concerns, which entails that the sense of pleasure associated with beauty is not mediated by knowledge (Carroll 2001, 25–27). Relatedly, for Kant, perhaps one of the most influential 18th-century proponents of the disinterested view, judgements of beauty can fall short of being pure, or free, if they are dependent, or involve classifying the object as falling within a kind and then considering the extent to which its function is achieved (Davies 2006, 225).[1] As Angela Breitenbach has summarised, 'Kant construes aesthetic judgments as involving the free play of imagination and understanding, and hence as essentially distinct from determinate cognition' (2018, 282).[2]

So, returning to the 18th-century experiments, because they were '*not* understood as controlled arrangements which test a theory, nature can be seen as expressing itself freely, authentically, in these experiments ...' (Rueger 1997, 317). Towards the end of the 18th century, however, the inductively oriented approach gave way to the hypothetic-deductive methodology. As Glenn Parsons and Rueger (2000) highlight, this correlated with a shift in the way that experiments were aesthetically appreciated: from the appreciation of nature that the experiment revealed to the design of the experiment itself.

With this change in methodology came a change in understanding: experiments were instead understood as controlled arrangements which test a theory. Indeed, control became key to obtain stable, repeatable results that could provide data to inform and confirm theories. This in turn bolstered the epistemic status of experiments compared to other practices in science. Adrian Currie and Arnon Levy for instance have proposed that, along with other factors, control underwrites 'experimentation's privileged status vis-à-vis confirmation' (2019, 1074). Accordingly, the attention shifted from *what* an experiment shows to *how* an experiment relates to a theoretical claim. With this, an experiment is aesthetically valuable, as Parsons and Rueger elaborate:

> ...because it shows 'aptness' in the relation of result and tools, of plan and success; it is a beautiful artefact, a manifestation of human ingenuity, an instrument optimally suited to achieve its purpose. What is appreciated is, for instance, the simplicity of the arrangement, its economy or its ability to unify several tasks in one display. (2000, 411–412)

In sum: 'many experiments that are considered beautiful are praised for their economy, simplicity and elegance' (Ivanova 2021, 4).

It has been posited that such aesthetic properties may also relate to the epistemic success of an experiment. Take the simplicity of an arrangement, as described in relation to experiments that prove white synthesis – where refracted light split into its multicoloured parts is mixed back into white again. Olaf L. Müller has articulated "aesthetic discomfort" with Isaac Newton's late 17th-century experiment, which involved placing a convex lens into the set-up (2010, 154–155). While veracious, in Müller's words the set-up:

> requires quite a lot of fumbling. The lens must be placed just so that the desired white spot can be seen without disturbances from the other colours; and that only works if the distances between prism, lens, and screen are adjusted to each other and to the properties of the lens.
> (2010, 155)

This, Müller, argues reflects not only a superficial imperfection but 'corresponds to a deeper aesthetic imperfection of the experiment. With the convex lens, an entirely new element comes into play…' (2010, 155). However, Müller goes on to discuss John Theophilus Desaguliers' early 18th-century version of white synthesis, which does not involve a convex lens but instead exploits the fact that some of the light rays return along the same path they came, as simpler and delivers a clear phenomenon to both the eye and intellect. For Müller, a simpler experiment is 'thereby [a] more beautiful experiment' (2010, 155).

One could argue that this is not really an aesthetic, but a "masked" epistemic assessment (Todd 2008). However, importantly, this assessment is being made in relation to how the form of the experiment is conveying the epistemic content. Furthermore, Müller's assessment that in this case, the simpler experiment is the more beautiful coheres with Parsons and Rueger's proposal that "operational simplicity" – 'an experiment is operationally simple iff it consists of few operations or experimental steps' (2000, 413) – is systematically connected to the epistemic success of experiments. Essentially, the simpler the set-up, the greater the reliability of the results as there is just less to go wrong or give misleading results.

Positive aesthetic and epistemic valuations also correspond to judgements of economy, sharpness, and elegance in experiments.[3] One such experiment that is highly praised in these respects is the 1958 Meselson-Stahl experiment that supported the theory that DNA replicates through a semi-conservative mechanism (Figure 2.1). As Ivanova recounts, the experimenters:

> fed bacteria nutrients containing heavy nitrogen isotope. Through metabolising, the nitrogen was slowly incorporated into the bacterial molecules. Meselson and Stahl continued the process until the genetic material of the bacteria became heavy. They then fed the

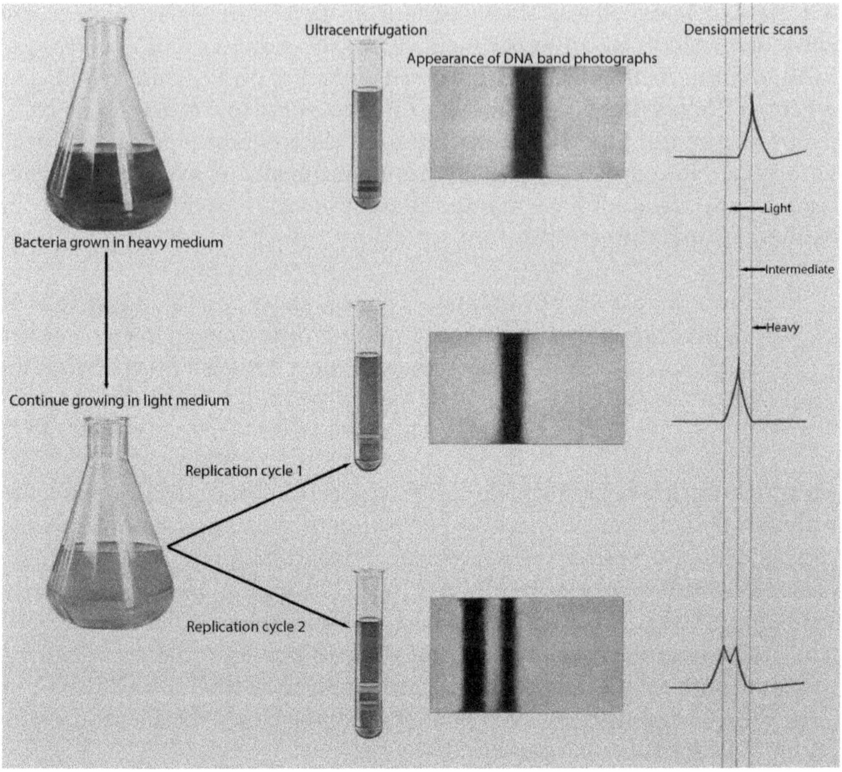

Figure 2.1 Drawing by the author illustrating the Meselson-Stahl experiment.

> bacteria light nitrogen and studied the genetic material of the bacteria through the next generations. They used a new method, ultracentrifugation, which allowed them to separate light from heavy genetic material. The ratios they obtained of light heavy and hybrid DNA were compatible with the semi-conservative replication and not with the conservative or dispersive model of replication.
>
> (2021, 4–5)

The design and results of the experiment are frequently described as beautiful.

For Frederic Holmes, 'the beauty was concentrated on the sharpness and significance of the outcome' (1997, 93). However, not only were the results aesthetically pleasing, but also the economy demonstrated in the design of the experiment and how this could be communicated: 'The result can be expressed in an elegantly simple form, and the central experimental operations can also be described simply. Underlying each experimental operation, however, are more complex foundations' (Holmes 1997, 100). That is, there is a wide range and depth of scientific knowledge required to 'fully explain the experiment' so that 'the simplicity of the original experiment, built on complexity, is, not elementary, but profound' (Holmes 1997, 100). Further to this, Ernst Peter Fischer has written of the beauty of 'making the genetic material physically heavier without changing it chemically' (quoted in Ivanova 2021, 5), while Ivanova has written that 'the use of centrifugation was innovative in this very step in their design was regarded as beautiful' (2021, 5). Indeed, as Ivanova notes, the use of innovative materials and original, unexpected steps in an experiment can also contribute to the judgement of beauty (2022). Similar displays of creativity and imagination can however lead to very different aesthetic experiences in the realm of experimental art.

2.2 Experimental Art

The 20th century saw avant-garde artists, including Dadists and conceptual artists, trialling the 'dissolution of boundaries in the arts' (Stallschus 2013, 15) and testing the boundaries of art itself, and, by extension, social, political, and philosophical ideas that permeated art. Hence, the approach of artists developing innovative, destabilising approaches to art and art-making tends to be termed "experimental".[4] As with scientific experimentation, there was a shift in the appreciative focus of the work from the content to how the results were realised.

It was a standard presumption from the 18th to the early 20th century that art should possess beauty (Danto 2003, xiv). However, this was a particular kind of beauty narrowly focused on the senses and corresponding to the disinterested pleasure discussed in the previous section. It was arguably this transposition of theories of disinterested, sensuous pleasure onto art (Carroll 2001, 23) that led 20th-century avant-garde artists to reject beauty. As Elisabeth Schellekens has explained, for artists who took a conceptual approach, 'modernism's loyalty to the notions of beauty, aesthetic sensation, and pleasing form, represented a commitment to obsolete artistic axioms' (2007, 72). Moreover, according to Arthur Danto, considering that theorists like Kant held beauty as a symbol of morality, 'it was the moral weight that was assigned to beauty that helps us understand why the first generation of the twentieth-century avant-garde found it so urgent to dislodge beauty from its mistaken place in the philosophy of art' (2002, 39).

Danto explained that Dada, an absurdist art movement that reacted to the horrors of World War I, exemplified what he termed the "Intractable Avant-Garde" – it not only refuses to be found beautiful, but he argued that the works are 'misperceived if perceived as beautiful. That is not its point or ambition' (2002, 47). For example, many have tried to analyse Duchamp's *Fountain* (Figure 2.2), a "readymade" sculpture consisting of an industrially made urinal that was placed on a plinth, in formalist terms and reported the smooth curves and white glazed surface of the

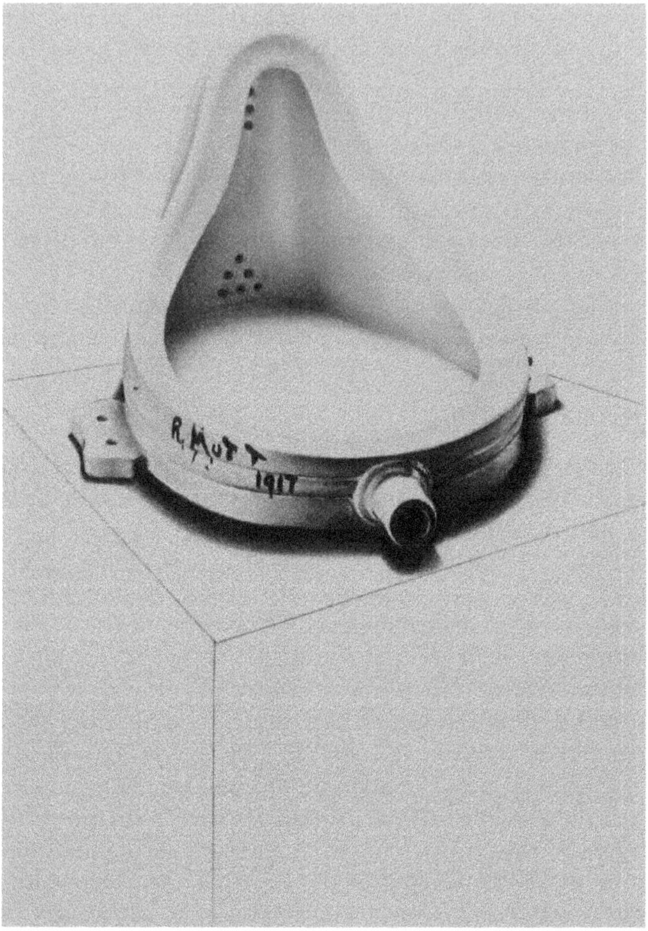

Figure 2.2 Drawing after Marcel Duchamp *Fountain* 1917 by the author.

urinal as beautiful. However, in Danto's view, any such beauty is incidental and not part of the work's meaning (2003, 9). To see whether the work itself is beautiful, Danto proposed that 'we have to identify the meaning of the work as given by thought' (2003, 95). As Danto highlighted, Duchamp had stated that the choice of readymades was based on a reaction of visual indifference (2003, 10). Rather than alerting us to the formal beauties of everyday objects, Duchamp sought with these simple, carefully chosen pieces 'to exemplify the most radical dissociation of aesthetics from art' (Danto 2003, 94). Through such economic means, avant-garde artists like Duchamp swiftly and often humorously dislodged beauty from art – concepts that had seemingly become synonymous for many. With this, artists could experiment with a host of social, political, and philosophical principles that permeated art.

Allan Kaprow, one of the pioneers of performance art, said that experimentation is the "testing or trial of a principle" (as quoted in Lapointe 2015). The experimentation seen in avant-garde art, however, was not necessarily the strictly controlled testing of a hypothesis, but as Andrew Pickering (2016) has proposed, open-ended like some forms of scientific experimentation. Artistic experiments, however, typically delivered further questions, rather than answers (Douglas and Gulari 2015; Lapointe 2015) that could be neatly prised from the results as in the most highly praised scientific experiments. For instance, in relation to John Cage's works, Pickering has said: 'The "experiment" in this class of works is to find out how the work itself will perform in time. And, on, the other side of the coin, the experiment lies in relinquishing, in part at least, the authorial control that is the mark of western mastery' (2016, 3). In these respects, the works were economically designed to test how, in contrast to the usual association of artworks with finished artefacts (Pickering 2016, 3), unfinished, dynamic systems evolved and what would happen when authorial control was reduced. Works, such as Cage's *4'33"* (1952), while simple were successful in destabilising the long-held expectations that artworks were finished artefacts fully under the control of their authors, but raised many thorny questions about the ontology of artworks for example.

These experimental themes were common in avant-garde performance arts, such as Happenings, where the aim was to explore the porous boundaries between art and life and set-up situations in which the audience "completed" the work (Douglas and Gulari 2015, 401). In producing these kinds of works, artists sought to go even further than Dadaists and not only produce indifferent or displeasing forms but actually "dematerialise" art. In doing so, they also sought to challenge 'the traditional emphasis on sensory pleasure and beauty' and this was frequently achieved by replacing 'illustrative representation' with 'semantic representation' (Goldie and Schellekens 2007, xii–xiii). Entire artworks, once great sensuous affairs, came to consist of a line of text for example. Thus, these works expressed their ideas simply and, in many cases, economically by the minimal means that were used to embody them. Essentially, artists rejected the aesthetic in order to elevate the cognitive (Schellekens 2007, 73). Although some attempts (by figures including art critic Dave Hickey for example) were made to revive beauty in the arts in the 1990s, as Danto has noted, what followed was less the pursuit of beauty than the idea of beauty: 'my sense is that it was exciting less because of beauty itself, than because beauty was proxy for something that had almost disappeared from most of one's encounters with art, namely enjoyment and pleasure' (2003, 8). So, despite sharing simple and economic designs, the scientific experiments are highly praised as beautiful, while the artistic experiments are not. Are there any conceptions of beauty that can accommodate why the simplicity and economy exhibited in scientific experimentation, but not art tends to produce a sense of pleasure in the appreciation of these practices?

2.3 Accounting for Aesthetic Judgements of Artistic and Scientific Experiments

As we have seen, notions of disinterested, sensuous pleasure have tended to dominate conceptions of beauty. While a Kantian notion of disinterestedness does not seem particularly amenable to aesthetic experiences of scientific or artistic experimentation, Breitenbach (2018) and Diarmuid Costello (2007) have each tried to demonstrate that it can plausibly accommodate such practices.

Breitenbach has argued it would be mistaken to rule out the beauty of science based on Kant's aesthetics on the grounds that scientific artefacts can elicit the kind of reflection required if they are the products of creative intellectual activity – something that is an open-ended pursuit – 'that makes possible a distinctive aesthetic experience in science' (2018, 282). As Breitenbach has highlighted, Kant does promote entities that deal with concepts and ideas, such as poetry, as key examples of beauty (2018, 289). In such cases, Breitenbach has explained: 'The expressions of concepts and judgments brought about by the capacity of genius leave room for the free play of the faculties and, hence, for aesthetic pleasure. Kant thus accounts for judgments of beauty that involve reflection on conceptual representations' (2018, 290).

With this, Breitenbach has used a Kantian lens to examine a range of scientific artefacts, including representations of scientific concepts and experiments. In relation to the latter, Breitenbach used the case of Louis Pasteur's manual separation of mirror-image molecular forms of tartaric acid, which is often regarded as one of the most beautiful experiments in the history of chemistry (2018, 294):

> Pasteur's conceptually simple experiment shows, in a carefully executed way, that chiral molecules – molecules with a right- or left-handed orientation – are optically active. They rotate polarized light unless there are mirror-image molecular groupings in the mix that cancel the effect. Pasteur's experiment reveals clearly and unambiguously this important truth in chemistry.
>
> (Breitenbach 2018, 294)

As Breitenbach proposed, 'the beauty of the experiment lies in its capacity to show in one glimpse a universal chemical principle underlying a whole range of phenomena' (2018, 294–295). However, it by no means exhausts the imagination as a result, leaving 'indefinite room for the imagination to contemplate the implications of this principle and to run through an inexhaustible string of associated representations. It expresses a universal idea by stimulating thought beyond the particular case represented and by procuring images of the whole of nature for which the principle holds' (Breitenbach 2018, 295). So, on the Kantian account, Breitenbach has suggested: 'we can appreciate Pasteur's experiment as beautiful because it presents a scientific idea in a way that leaves room for free play of the faculties' (2018, 294–295).

Likewise, what Costello has emphasised from Kant's 'account of works of art as expressions of aesthetic ideas is his stress on the *imaginative engagement with ideas* that works of art induce in the spectator, far removed from the astringent formalism generally attributed to the third *Critique* as a reception aesthetic' (2007, 104). Turning to conceptual art with this framing, Costello has proposed that the success of these works lies with the fact that the material embodiment of the ideas they explore can trigger a range of associations and thoughts (2007, 111). As such, Costello has suggested that conceptual artworks 'may still be called aesthetic, in Kant's terms, by virtue of triggering a wealth of imaginative associations through the indirect presentation of an idea in sensible form' (2007, 111–112). Costello has noted though, that the form of some conceptual works is limited to a line of text – from which he has suggested that 'their lack of form enables them to fulfil, in a pronounced way, Kant's claim that aesthetic ideas 'emulate the example of reason in striving for a maximum', that is, for a completeness that experience does not afford' (2007, 112).

Although Breitenbach and Costello are right to suggest that artistic and scientific experiments can present ideas in ways that productively trigger the imagination, there are other aspects of the reported experiences of these entities that their accounts fail to accommodate. For one thing, although conceptual works can provoke a wealth of imaginative associations, they still do not seem to yield the aesthetic pleasure that should, as per a Kantian account, follow. Furthermore, Breitenbach seems to focus on the effects of appreciating the result of Pasteur's "carefully executed" experiment. However, as this description highlights, and as per the examples from Section 2.1, such a focus neglects the fact that the design of scientific experiments seems central to evaluating these practices, not only the results. As Ivanova has written, descriptions of 'aptness for purpose' are common to praise many beautiful experiments (2021, 4). These factors can be taken to suggest that while aspects of the experiences under discussion can be accounted for with a disinterested approach, we ought to look to an altogether different conception of beauty to fully explain the responses to the artistic and scientific experiments under discussion.

2.4 Functional Beauty

Although notions of disinterested pleasure have tended to dominate the aesthetic tradition, another approach is to factor in the function of an object. Accounts of FB have ancient origins and since this time, as

Andrea Sauchelli has outlined, two broad kinds of accounts of FB have emerged: external and internal (2013, 41). Externalist accounts have it that considerations of function are external to judgements of beauty, so that 'the functional appropriateness of an object is a necessary but not sufficient condition for such an object to be considered beautiful' (Sauchelli 2013, 42). Typically, such an approach entails that for something, which has a specific function, to be judged as beautiful, it must fulfil its function. Should an ornate portable object that is meant to produce flames fail to ignite anything then the object would not be judged beautiful as a lighter. This kind of approach can be seen in Kant's notion of dependent beauty.

While externalist accounts hold that functional considerations constrain, or only indirectly or negatively contribute to aesthetic judgements, internalist accounts suggest that 'an object can also be judged as beautiful by virtue of the way it looks in relation to its function' (Sauchelli 2013, 42). Generally, internalist theories admit that FB is one kind of beauty, and that beauty may not be solely explained in terms of utility. Although, as Sauchelli highlighted, within this family of theories, there are further distinctions (2013, 43). For some, such as Stephen Davies (2006), the thing in question is functionally beautiful when its aesthetic properties contribute positively to it performing its intended principal function. As Davies has suggested: 'A beautiful chair is one having features that make it graceful and stylish and, at the same time, comfortable to sit on, stable and supportive of the back' (2006, 237). For others, however, such as Parsons and Allen Carlson (2008), a thing's aesthetic qualities emerge from its function. It is this latter kind that we will focus on here, given that the positive evaluation of artistic and scientific experiments is frequently explained in terms of how the results are realised by the design of the forms that realised them.

2.4.1 Parsons and Carlson's Internalist Account of Functional Beauty

To get their account off the ground, Parsons and Carlson propose solutions to the "problem of translation", or how awareness of an object's function can alter the aesthetic properties we perceive it to have (2008, 46), and the "problem of indeterminacy", concerning how to pinpoint the function of an object (2008, 49). For the latter problem, they suggest a selected effects account: 'X has a proper function F if and only if Xs currently exist because, in the recent past, ancestors of X were successful in meeting some need or want in the marketplace because they performed F, leading to manufacture and distribution of Xs' (2008, 75). This entails that the functions of objects can alter over time – a barn can become a house for example. For the former problem, they draw upon Kendall Walton's categories of art argument (1970), according to which the aesthetic properties of a work depend upon knowledge about the category the work belongs to. For any category, there are features which are standard, variable or contra-standard. Pigment applied to a support is standard for a Dutch still life painting for example, while the size of the support is variable and three-dimensionality is contra-standard. Applying this approach to functional categories, Parsons and Carlson propose that things might 'look fit' when 'the functional category employed to perceive them causes us to see those things as having no contra-standard features at all and as having, to a high degree, certain variable features that are indicative of functionality' (2008, 95–96). Should an object possess only standard features however, then it may appear elegant or graceful, while, in some cases, an object may be functional but possess contra-standard features that give rise to a pleasing visual tension (Parsons and Carlson 2008, 94).[5]

2.4.2 *Revised Functional Beauty*

While Parsons and Carlson's account is enlightening, as Panos Paris has highlighted, it still primarily concerns objects that are available through visual or aural perception and not, for example, scientific theories and mathematical proofs (2020, 519).[6] Furthermore (and importantly for our art cases), there are many counter examples, such as objects that appear well-formed for their functions, like bins and urinals, yet are aesthetically-indifferent (Paris 2020, 520). To overcome these issues, Paris has proposed the following:

> Revised FB (RFB) = If an object, O, is (1) well-formed for its function(s), and (2) pleases most competent judges in so far as it is experienced (in perception or contemplation) as (1), then O is (functionally) beautiful.
>
> (2020, 521)

Paris clarified that: (1) should be understood in terms of 'proper function' and (2) incorporates not only knowledge, 'but also a broader notion of expertise, in line with Hume's notion of a 'true judge', possessing '[s]trong sense, ... delicate sentiment, improved by practice, ... comparison, and cleared of all prejudice' [1777: 241]' (2020, 521). With some further clarifications, the RFB account, I propose, best explains the range of responses to the artistic and scientific experiments under discussion.

2.4.3 *The Issue of Proper Function*

We might be sceptical about how the notion of proper function applies to artefacts like artworks. Before addressing this issue, however, we need to clear up what notion of proper function is most appropriate for the RFB account. While Paris stated that he does not wish to confine

RFB to any particular account of proper function, he has noted 'perhaps no single conception applies across all objects that can intelligibly be said to have proper functions' (2020, 521). Indeed, while an admirable quality of Parsons and Carlson's selected effects theory is that it tries to take both biological and artefactual entities in under one conception of proper function, this has been a consistent source of criticism for their account (Shiner 2009; Davies 2010; Price 2010; Gilmore 2011; Stecker 2011).

One notable objection is that of artefacts with novel functions – the possession of which might not be explained in terms of corresponding capacities being present in earlier artefacts (Gilmore 2011, 297). Davies (2010), Robert Stecker (2011, 2019), and Jonathan Gilmore (2011) have each suggested that some reference to intentions is needed to account for the function of these objects. Gilmore, for instance, has proposed that: 'a more plausible account of artifactual function would identify it in the intention with which an artifact is created, whatever the subsequent history of that artifact's kind turns out to be' (2011, 297). As he acknowledges, this may come at the cost of a concept of function that unifies artefacts and biological entities (Gilmore 2011, 297). Davies, meanwhile, has proposed that: 'we can accept that market endorsement is necessary to an item's proper function while also maintaining a moderate intentionalism that makes the designer's purpose no less relevant to the outcome, at least in the initial phase' (2010, 323). Along similar lines, Stecker has suggested that a theory that combines intention with the way individuals belonging to the kind are used might, in a messy way, account for why functional artefactual kinds are created, why they become public kinds, and help us to identify the current primary function of a kind (2019, 24).

Although it would be impressive to have an account of proper function that accommodates both biological and artefactual entities in one formulation, it would be better to acknowledge that there are some important differences in how these entities are created, given that there may be no precedent for the function of an artefact. Accordingly, like Davies, I propose that a moderate intentionalism ought to be maintained when ascertaining the proper function of artefacts, including items of furniture, buildings, and scientific experiments. This kind of approach, which I will only be able to leave in schematic terms here, enables us to account for why objects are created, as well as why they may be reproduced and eventually alter over time.

With this, it is relatively straightforward to pin down the proper functions of different kinds of scientific experiments. As Ivanova has explained, experiments are performed with a plurality of aims – some are 'to confirm a hypothesis or discover the existence of a particle or phenomena' (2022, 13) while others 'include the exploration of new domains and uncovering phenomena that need not be accommodated within an existing theoretical framework' (2022, 13). We should also note here that Paris wrote of an object as being functionally beautiful if it is well-formed for its function(*s*). Multifunctional objects, Paris has suggested, can be handled differently, depending on the case at hand – they might be seen in light of a unitary function (realising a number of functions at once) or appreciated in a *pro tanto* fashion, or their form might be considered in light of functions X, Y, etc. (2020, 517, n.2).[7] Given the earlier examples where beauty was attributed to both the design and results, it is likely that a single experiment may have more than one proper function – more on this later.

The proper function(s) of avant-garde arts have tended to be harder to pin down. As Parsons and Carlson point out, categories of religious and memorial art are quite plausibly connected to proper functions of inspiring awe or emotion (2008, 220, 224). Avant-garde art, however, developed out of an "autonomous" idea of art (Bernard 2018, 3) that has its origins in the 18th century (Parsons and Carlson 2008, 26–27). Nevertheless, as Davies posited, in his account of FB, such "autonomous artworks" arguably have the function of being pleasing when contemplated for their own sake (2006, 228). While, as a number of philosophers of art have noted, many avant-garde kinds of art are made so as to be self-reflexive – they explore the boundaries of art and the nature of art itself.

Parsons and Carlson express scepticism at the prospect of motivating such a function, questioning: 'What sort of non-aesthetic perceptual qualities, for instance, would be standard in a work that aims to stimulate thinking about art? If experimental art represents a new sort of tool, then, perhaps because of its newness, it is yet unclear how and to what extent Functional Beauty applies to it' (2008, 226–227). Although they pose this as a problem for their account of FB, it is not clear that they need to. For one thing, it seems that there is a set of perceptual qualities that experimental artworks share, namely works of this kind usually rely on arrangements of means and materials from everyday life, rather than the plastic arts. As Yuriko Saito has highlighted, 'twentieth century art introduced and experimented with different modes of appropriating everyday life, most famously with the use of ready-mades' (2021). Besides, with RFB, we need not be as concerned about perceptual qualities as Parsons and Carlson need to be with their account of FB. Rather than looking fit, Paris offered a sufficient condition for beauty whereby something is well-formed for its function(s) and is pleasingly experienced, either perceptually or in contemplation, as such. While conceptual artworks are neither identical with nor reducible to ideas, as Hanson has proposed contra to much contemporary philosophising on the subject, descriptions can acquaint you with *aspects of* a conceptual work (2015, 256). Accordingly, conceptual works can be apprehended, to a certain extent, in contemplation, as well as perception. All this is to say that there are plenty of options to explain how avant-garde artworks might be experienced as well-formed for their destabilising functions.

2.4.4 The Role of Competent Judges

In relation to the experience of objects, as Paris clarified for (2), the notion of a competent judge incorporates not only knowledge but also a broader notion of expertise in line with Hume's notion of a true judge. As Paris highlighted, 'What is important is that a 'competent judge' is an expert relative to a given domain' (2020, 521–522). This aspect of RFB may go some way to accounting for why typically the artistic and scientific experiments yield quite different experiences, despite sharing properties such as simplicity and economy. Notably, as the earlier discussions of the beauty of scientific experiments demonstrate – understanding what an experiment shows and how it does this are central to the sense of pleasure taken in appreciating it. Those who are experts in the scientific domains at hand are those who seem to praise the experiments in the highest terms for their beauty, something that (2) tracks. Many audiences encountering the avant-garde works, however, are not experts in this domain and, as in the case of *Fountain*, even some experts may miscategorise these works.[8] Such cases make it clear why considering the intentions of creators is also important in ascertaining the proper function(s) of artefacts like artworks.

It is likely that many audiences have viewed, and continue to mistakenly view, avant-garde art under an aesthetic functional kind, where, as Davies has suggested, the artworks arguably have the function of being pleasing when contemplated for their own sake.[9] There is plenty of evidence to suggest that formalist conceptions of art lingered on well into the 20th century (and probably even now). Engler, for instance, proposed in 1990 that in art 'beauty is considered predominately as an object of the senses' (1990, 33). While, as Costello (2007) has highlighted, in the mid-to-late 20th century the highly influential artworld figure Clement Greenberg interpreted, and in doing so distorted, Kant's theory of taste to underwrite his modernist theory. In doing so, Greenberg identified art with aesthetic experience and saw medium-specificity as the way to pursue aesthetic value (Costello 2007, 95). Greenberg's formalist agenda did not survive although, as Costello has argued, 'what Greenberg once valued is now roundly devalued, but what has *not* changed is the understanding of aesthetics underpinning his critics' position' (2007, 100).

Indeed, without this, many avant-garde artists may not have been driven to so resolutely pursue their anti-aesthetic agenda. Understood as vehicles for this agenda, avant-garde artworks are well-formed for their functions, including critiquing and challenging the nature of art, and experimenting with a host of social, political, and philosophical principles that permeated art, such as authorship, the divide between artist and audience, and gender roles. Moreover, there is evidence that they do please competent judges.[10] As Danto has noted of works like Yoko Ono's early "instruction work", anyone could perform '*Match Piece*, in which one strikes a match and watches it until the flame goes out. The dying of the light – what could be more poetic?' (2003, 22). It is important to note in this example, however, that the work also featured some attractive formal properties – who hasn't admired the glow of flames and the way they dance and change? This is not the case, as indicated earlier, for very many avant-garde works.

For further evidence of this, take Mary Kelly's *Post-Partum Document* (1974) (interestingly, while an experimental artwork, the form of the work actually models a scientific mode of observation).[11] As a work that cuts through the entire history of art with the simple gesture of typewritten observations and the minimal visual traces of her son's development on the support (stained nappy liners), the work was extremely controversial, but sharp and economic, as a radical reinterpretation of the representation of motherhood. There is not much that is pleasant to be found in dried excrement and this is a really important point of the work. It subverts the artistic tradition of sentimentalising motherhood (Goodwin 2018) – dirty nappies are hardly a standard feature of traditional mother and child scenes, and the shock of this very real, rather unpleasant aspect of parenting very forcefully makes this happen. Therefore, while it is beautiful in a functional sense – for those who understand its place in the history of art – it fails to please the senses and this in fact is quite crucial for its point. This, I propose, is why, even when a work is found to be functionally beautiful, the overall experience of appreciating it may not be a pleasurable one.

2.5 The Distinction between Formal and Functional Beauty

As those who advance accounts of FB have stressed, it is likely to be only one kind of beauty, one that is grounded in the function of an object (Paris 2020, 518), and FB is not always equivalent to formal beauty (Davies 2006, 238). Parsons and Carlson noted that 'the pig's snout may displease us with its texture, dirty appearance, and foul smell, but nonetheless contain a beauty, albeit one that is obscured, in its aptness of form to function' (2008, 16). Indeed, both Parsons and Carlson, and Paris construe their versions of FB in a *pro tanto* fashion so that 'an object can be functionally beautiful in so far as it looks well-formed for its function(s), although also ugly in terms of its shape or colour' (Paris 2020, 518). Objects can also be ugly insofar as they are well-formed for immoral purposes (Paris 2020, 526). Furthermore, it may be the case that not all aspects of an item contribute to its FB – as Roberts has highlighted, an object may feel but not look functionally beautiful (2022, 58, n.13). Accordingly, it is likely that avant-garde artworks often fail to be pleasurably appreciated because, even if well-formed for their purpose(s), the artefacts are, in other respects, aesthetically indifferent or unpleasant. Given the examples under discussion here, the following will focus primarily on how formal properties contribute to the experience of appreciating artefacts.[12]

Some formal properties, such as regular shapes, smooth textures, and fractal forms (Carafoli 2016) evidence would suggest, tend to be more pleasing than others due to neural factors, including processing fluency (Todd 2020). That is, our brains are able to process these features more fluently, a feeling, which we can experience (Arcangeli and Dokic 2020, 106). Some artefacts may even be experienced as perceptually disfluent but conceptually fluent, such as Bridget Riley's op-art paintings (Arcangeli and Dokic 2020, 118). In a similar vein, the preference among scientists for properties like simplicity may also reflect the limits of human cognitive abilities or the limits of plasticity (Montano 2013). To use Margherita Arcangeli and Jérôme Dokic's example: 'a long and complex sentence feels difficult to parse and to understand, in contrast to a short and simple one' (2020, 115). Accordingly, forms that are easier for our brains to process are potentially more likely to be experienced as pleasurable. It seems likely then, that for the appreciation of an artefact to yield an overall pleasurable experience, it needs to have properties that, at the very least, do not detract from its FB and better still are also themselves conducive to being experienced as pleasurable.

Indeed, one common objection raised to FB was the issue of aesthetically indifferent objects. As Stecker remarked: 'the fact is that many artifacts are aesthetically indifferent even though they are well designed to fulfil their function or functions on whichever conception of function that is relevant to appreciation' (2019, 27). Consequently, Stecker goes on to propose that 'design features that have formal aesthetic interest are necessary for ordinary artifacts to have aesthetic value' (2019, 28). This does not mean that these design features will always enhance an object's primary function. To illustrate this, Stecker used the example of plates that have a pleasing visual pattern painted on them (2019, 29). Such additions do not obviously make the plates better for arranging and consuming food but do plausibly enhance the experience of a meal. In other cases, however, particular artefact types can provide opportunities to present aesthetically pleasing appearances – as in the sculpted base of a table lamp, for example, which can enhance the experience of a gathering in the room illuminated by the lamp (Stecker 2019, 29–30). In these cases, this may be a secondary function that emerges from the primary one but 'the aesthetic features do not strictly have to enhance the *primary* function of an artefact to contribute to its functional value' (2019, 30), Stecker has proposed.

Paris' second clause, as Roberts has outlined, for RFB avoids the counterexample of attributing beauty too widely (2022, 53). Most competent judges are not pleased by well-designed vehicles for disposing of waste for example. However, it does seem that Stecker's account might explain why some well-designed artefacts are not apt to please even competent judges if design features must instantiate formal aesthetic interest. That is, to merit an overall positive aesthetic assessment, it is also necessary for artefacts to instantiate formal aesthetic interest. This is something that many of the scientific experiments do achieve with well-made but also formally interesting instruments or methods for example. Furthermore, the phenomena under investigation, as the Meselson-Stahl or white synthesis experiments show, often have formal properties that are found to be very pleasing and contribute to the pleasurable appreciation of the design of these experiments and their results.

By contrast, while many works of experimental art might be well-formed for their functions, this state is achieved by artists employing formal properties that are aesthetically indifferent, or even unpleasant. As Danto reminds us, the point of the ground-breaking work *Fountain* is that it is not beautiful: an aesthetically indifferent, utilitarian, everyday object was chosen for precisely this provocative reason and is central to the artistic value of the work. Likewise, as Costello has written: 'The fixation on lists, diagrams, and maps, and the cheap 'xerox aesthetic' to which many early commentators on Conceptual Art drew attention is just that, *an aesthetic* – if not an especially gratifying one, sensuously' (2007, 113). Hence, while these works are well-designed to discharge their functions, and may indeed engender a positive evaluation on that level, this is frequently achieved at the cost of formal beauty. Acknowledging the different species of beauty and the interaction between them in making aesthetic evaluations, as the foregoing demonstrates, means that we can account for the responses to the artistic and scientific experiments under discussion.

2.6 Conclusion

As we have seen, despite sharing simple and economic designs, scientific experiments are much more likely to attract praise for being beautiful than artistic experiments. Factoring in FB gives us the advantage of accounting for these differing responses. It also has the advantage of explaining the relationship between aesthetic and artistic values in the experimental artworks, and the close relationship between aesthetic and epistemic values in the scientific experiments. As I highlighted earlier, scepticism has been expressed at the prospect of genuinely aesthetic judgements in science. Yet, in common to the discussion of the scientific experiments outlined earlier was the appreciation of instruments optimally suited to achieve their purpose; simple, economic, or elegant setups; and aptness for purpose. The foregoing makes it manifestly clear that in many experimental cases, considerations of practical utility are not themselves aesthetic considerations (Todd 2008, 74), but that aesthetic judgements relate to the way in which the epistemic content of experiments is conveyed (Todd 2008, 77). Accordingly, we have grounds to argue that scepticism about the role of beauty in scientific progress is unwarranted in cases where judgements of beauty are made in relation to how well-designed an experiment is to fulfil its function. After all, surely it is rational to try and optimise the design of experiments. Doing so refines experimental practices to better realise their epistemic function(s), which may vary depending on the experimental aims.

Generally, though, the epistemic functions of scientific experiments seem to ideally include not only the reliable production of epistemically useful data, but the production of this in a way that reveals the results (whether they lead to a new discovery, provide further evidence for, or dismiss, a theory, or explore new directions), as unambiguously as possible, or sharply and clearly. In fact, to return to the earlier point about multifunctional entities, it seems that scientific experiments are valued most highly in light of a unitary function, which corresponds to the value found in experiments that exhibit economy through elegant design. Take the Meselson-Stahl experiment – plausibly, the economy of this experiment also relates to its epistemic success, considering the design of the experiment made it possible to simultaneously confirm one hypothesis and discredit others. FB in much experimental art, by contrast, seems to be most appropriately appreciated in a *pro tanto* fashion. That is, insofar as the works are well-formed to disrupt and destabilise common conceptions of art, including social, political, and philosophical ideas that permeate art, they are functionally beautiful, but ugly in terms of their formal properties. Experimental artworks, in keeping with the aims of avant-garde artists, tend to deliver further questions rather than clear or sharp results, and indeed, the more disruptive they are in this respect, the more successful the work tends to be. Accordingly, the displeasing properties may also be present in appreciating the results as the questions for art tend to be messy and complex.

Both artistic and scientific experiments can exhibit sharpness, simplicity, and economy in their designs, although whether these properties are manifested in the results may vastly differ between these domains. Nevertheless, the presence of these properties in the designs of artistic and scientific experiments may not only track, but actually contribute to their success. Hence, these experiments may be experienced as functionally beautiful if they are well-formed for their functions. As the foregoing demonstrates, however, for the appreciation of these experiments to be an overall pleasurable experience, formal beauty is also important and so, despite having certain properties in common to their designs, artistic and scientific experiments tend to give rise to quite different aesthetic experiences – the beauty of the account advanced here helps us to explain why this might be.

Notes

1 See Davies (2006) for a critique of the notion of free beauty.
2 Breitenbach has further specified that, for Kant: 'beautiful objects are purposive since, by reflecting on the form of such objects, our cognitive faculties are "unintentionally brought into accord with one another." More specifically, since we do not become conscious of this fit by subsuming the object under determinate concepts, but through a feeling of pleasure, aesthetic judgment has a distinctive subjective character. Moreover, since this pleasure is related to a harmony of capacities common to all, and not dependent on the satisfaction of any particular idiosyncratic desires, aesthetic pleasure can be imputed to all. This is why Kant maintains that the free play of imagination and understanding is associated with a feeling that has universal validity' (2018, 288).
3 Elegance, according to Engler, 'is an aesthetic concept which is often used in the theory of beauty and has the meaning of: "chosen skilfully or carefully"' (1990, 30).
4 We could also discuss artistic experimentation in terms of the manipulation of materials or trialling of new techniques, and new methods of representation (Bernard 2018, 3). However, given my focus here, I will concentrate on the approach of experimental artists who sought to test the boundaries of the arts and art itself, given the disavowal of beauty in doing so.
5 See Davies (2010) and De Clerq (2013) for difficulties with, and objections to, using the notions of standard, variable, and contra-standard features in relation to function.
6 Although see Bueno (2009) for a way to potentially extend it to account for beauty in maths.
7 Artefacts can also have accidental or external functions. See Gilmore (2011) for a discussion of these and also De Clerq (2013) for an account of the hierarchy of various functions that objects might possess.
8 This was sometimes the case even for those close to Duchamp – as Danto outlined: 'It was Duchamp's patron, Walter Arensberg, who thought – or pretended to think – that disclosing beauty was the point of *Fountain…*' (2003, 9).
9 As Parsons has highlighted, in a paper exploring the role of form and function in forming judgements of beauty in chemistry, appreciating this kind of beauty requires specialized knowledge – something which is true for appreciating art as much as science (2012, 579).
10 Drawing upon personal experience, I once taught a confirmed group of aesthetes on a short art history course for adults about Neo-Dada. I was very intrigued when a few members of the group came up to me at the end of the session to tell me that, after years of dismissing the works as unappealing, they had a positive experience in engaging with the works, having learnt about why these artists challenged the boundaries between art and everyday life.
11 See also Ruffa (2005) for an account of Salvador Dali's parody and challenge to the model of scientific observation.
12 There are, however, plausible cases of avant-garde art which contain ethically dubious aspects that may also detract from pleasurable appreciation, even if the works might be well-formed for their function(s). For more on the interaction of moral and aesthetic considerations in appreciating artworks, see Dixon (2019).

Bibliography

Arcangeli, M. and Dokic, J. (2020) 'A Plea for the Sublime in Science', in S. French and M. Ivanova (eds.) *The Aesthetics of Science; Beauty, Imagination and Understanding*. New York: Routledge, pp. 104–124.

Bernard, C. (2018) 'In the Turbine of Experimentation: Tate Modern and the New (?)Rationale of Collective Performance', *Angles*, 6, 1–16.
Breitenbach, A. (2018) 'The Beauty of Science without the Science of Beauty: Kant and the Rationalists on the Aesthetics of Cognition', *Journal of the History of Philosophy*, 56(2), 281–304.
Bueno, O. (2009) 'Functional Beauty: Some Applications, Some Worries', *Philosophical Books*, 50(1), 47–54.
Carafoli, E. (2016) 'The Creativity Process: Freedom and Constraints', *Rendiconti Lincei*, 27, 413–425.
Carroll, N. (2001) *Beyond Aesthetics: Philosophical Essays*. Cambridge: Cambridge University Press.
Costello, D. (2007) 'Kant after LeWitt: Towards an Aesthetics of Conceptual Art', in P. Goldie and E. Schellekens (eds.) *Philosophy and Conceptual Art*. Oxford: Oxford University Press, pp. 92–115.
Currie, A. and Levy, A. (2019) 'Why Experiments Matter', *Inquiry*, 62(9–10), 1066–1090.
Danto, A. C. (2002) 'The Abuse of Beauty', *Daedalus*, 131(4), 35–56.
Danto, A. C. (2003) *The Abuse of Beauty: Aesthetics and the Concept of Art*. Chicago, IL: Open Court.
Davies, S. (2006) 'Aesthetic Judgements, Artworks and Functional Beauty', *The Philosophical Quarterly*, 56(2), 224–241.
Davies, S. (2010) 'Functional Beauty Examined', *Canadian Journal of Philosophy*, 40(2), 315–332.
De Clerq, R. (2013) 'Reflections on a Sofa Bed: Functional Beauty and Looking Fit', *The Journal of Aesthetic Education*, 47(2), 35–48.
Dixon, D. (2019). 'Conflicted Art: How to Approach Works by Morally Bad Artists', *Art Aesthetics Magazine*. https://www.artaesthetics.net/publications/2019/5/26/conflicted-art-how-to-approach-works-by-morally-bad-artists.
Douglas, A. and Gulari, M. N. (2015) 'Understanding Experimentation as Improvisation in Arts Research', *Qualitative Research Journal*, 15(4), 392–403.
Engler, G. (1990) 'Aesthetics in Science and in Art', *British Journal of Aesthetics*, 30(1), 24–34.
Gilmore, J. (2011) 'A Functional View of Artistic Evaluation', *Philosophical Studies*, 155, 289–305.
Goldie, P. and Schellekens, E. (2007) 'Introduction', *Philosophy and Conceptual Art*. Oxford: Oxford University Press, pp. ix–xxi.
Goodwin, A. (2018) 'Post-Partum Document. Documentation III: Analysed Markings and Diary Perspective Schema (Experimentum Mentis III: Weaning from the Dyad)', *Tate*. https://www.tate.org.uk/art/artworks/kelly-post-partum-document-documentation-iii-analysed-markings-and-diary-perspective-t03925.
Hanson, L. (2015) 'Conceptual Art and the Acquaintance Principle', *The Journal of Aesthetics and Art Criticism*, 73(3), 247–258.
Hickey, D. (1998) 'Enter the Dragon: On the Vernacular of Beauty', in B. Beckley and D. Shapiro (eds.) *Uncontrollable Beauty*. New York: Allworth Press, pp. 15–24.
Holmes, F. L. (1997) 'Beautiful Experiments in the Life Sciences', in A. I. Tauber (ed.) *The Elusive Synthesis: Aesthetics and Science*. Dordrecht: Kluwer Academic Publishers, pp. 83–102.

Ivanova, M. (2021) 'The Aesthetics of Scientific Experiments', *Philosophy Compass*, 16: e12730. https://doi.org/10.1111/phc3.12730.

Ivanova, M. (2022) 'What Is a Beautiful Experiment?', *Erkenntnis*. https://doi.org/10.1007/s10670-021-00509-3.

Lapointe, F. J. (2015) 'On the Role of Experimentation in Art (and Science)', *Journal of the New Media Caucus*. http://median.newmediacaucus.org/research-creation-explorations/6095-2/.

McMahon, J. A. (2005) 'Beauty', in B. Gaut and D. M. Lopes (eds.) *Routledge Companion to Aesthetics* 2nd Edition. New York: Routledge, pp. 307–319.

Montano, U. (2013) 'Beauty in Science: A New Model of the Role of Aesthetic Evaluations in Science', *European Journal for Philosophy of Science*, 3, 133–156.

Müller, O. L. (2010) 'Colour Spectral Counterpoints Case Study on Aesthetic Judgement in the Experimental Sciences', in I. Nussbaumer (ed.) *Restraint versus Intervention: Painting as Alignment*. Nuernberg: Verlag für moderne Kunst, pp. 150–171.

Paris, P. (2020) 'Functional Beauty, Pleasure, and Experience', *Australasian Journal of Philosophy*, 98(3), 516–530.

Parsons, G. (2012) 'The Aesthetics of Chemical Biology', *Current Opinion in Chemical Biology*, 16, 576–580.

Parsons, G. and Carlson, A. (2008) *Functional Beauty*. Oxford: Oxford University Press.

Parsons, G. and Rueger, A. (2000) 'The Epistemic Significance of Appreciating Experiments Aesthetically', *British Journal of Aesthetics*, 40(4), 407–423.

Pickering, A. (2016) 'Art, Science and Experiment', *MaHKUscript: Journal of Fine Art Research*, 1(1), 1–6.

Price, C. (2010) 'Functional Beauty', *British Journal of Aesthetics*, 50(2), 215–218.

Roberts, T. (2022) 'Feeling Fit for Function: Haptic Touch and Aesthetic Experience', *British Journal of Aesthetics*, 62(1), 49–61.

Rueger, A. (1997) 'Experiments, Nature and Aesthetic Experience in the Eighteenth Century', *British Journal of Aesthetics*, 37(4), 305–322.

Ruffa, A. (2005) 'Dalí's Surrealist Activities and the Model of Scientific Experimentation', *Papers of Surrealism*, 4, 1–14.

Saito, Y. (2021) 'Aesthetics of the Everyday', *The Stanford Encyclopedia of Philosophy*. https://plato.stanford.edu/archives/spr2021/entries/aesthetics-of-everyday/.

Sauchelli, A. (2013) 'Functional Beauty, Perception, and Aesthetic Judgements', *British Journal of Aesthetics*, 53(1), 41–53.

Schellekens, E. (2007) 'The Aesthetic Value of Ideas', in P. Goldie and E. Schellekens (eds.) *Philosophy and Conceptual Art*. Oxford: Oxford University Press, pp. 71–91.

Shiner, L. (2009) 'Functional Beauty by Parsons, Glenn and Allen Carlson', *The Journal of Aesthetics and Art Criticism*, 67(3), 341–343.

Stallschus, S. (2013) 'A Theory of Experimentation in Art? Reading Kubler's History of Art after Rheinberger's Experimental Systems', in M. Schwab, V. Anderson, P. de Assis, E. Bippus and H. Borgdorff (eds.) *Experimental Systems: Future Knowledge in Artistic Research*. Leuven: Leuven University Press, pp. 15–25.

Stecker, R. (2011) 'Review of Glenn Parsons and Allen Carlson's Functional Beauty', *The Philosophical Quarterly*, 61(243), 439–442.
Stecker, R. (2019) 'The Interactions of Function and Aesthetic Value in Artifacts', *Grazer Philosophische Studien*, 96(1), 19–36.
Todd, C. (2008) 'Unmasking the Truth Beneath the Beauty: Why the Supposed Aesthetic Judgements Made in Science May Not Be Aesthetic at All', *International Studies in the Philosophy of Science*, 11, 61–79.
Todd, C. (2020) 'Imagination, Aesthetic Feelings, and Scientific Reasoning', in S. French and M. Ivanova (eds.) *The Aesthetics of Science; Beauty, Imagination and Understanding*. New York: Routledge, pp. 63–85.
Walton, K. (1970) 'Categories of Art', *The Philosophical Review*, 79(3), 334–367.

3 Robert Boyle (1627–1691) and the Aesthetics of Chemical Experimentation

Alexander Wragge-Morley

3.1 Introduction

Robert Boyle (1627–1691) was perhaps the leading member of the early Royal Society of London, a scientific society founded in 1660 to promote an experimental approach to learning about natural things. The Society's position was not in itself a new one. Scientific practitioners had long mobilised empirical methods to figure out how nature worked, especially in the fields such as alchemy. Instead, what made the Society distinctive was its corporate insistence that, in most fields of inquiry, experimentalism should replace the varieties of deductive reasoning that then still enjoyed considerable prestige. Taking this position set the Society against a remarkably wide range of contemporary philosophers. On the one hand, the practice of deducing the truth *a priori* from a given premise was important to many thinkers attached to a university curriculum based, albeit with considerable modification, on the texts and principles left behind by the ancient Greek philosopher Aristotle. On the other hand, logical deduction remained important to many of those thinkers who explicitly rejected the Aristotelian inheritance. Supposedly modern philosophers such as Thomas Hobbes and René Descartes based their hypotheses about the physical world on what it seemed possible to figure out from first principles, using the powers of the mind alone (Dear 2009, 5–6, 81, 98, 127–131, 140–141).

As Steven Shapin and Simon Schaffer (1985) showed almost 40 years ago, the early Royal Society, therefore, owes much of its importance to its role in a dispute about the value of experiment. Boyle had at least two aims in mind when he pursued his now-famous programmes of experimentation in pneumatics and chemistry. It wasn't just that he wanted to show that the visible phenomena we take for granted depend on mechanical interactions between imperceptibly small particles of matter that he called corpuscles. In its outlines – if not the details – this was a position with which Hobbes and Descartes would have agreed. Rather, he wanted to show that the best way to vindicate this hypothesis was through experimentation. In other words, he sought to show that

DOI: 10.4324/9781003263920-4

experiments conducted in a laboratory could provide natural philosophers with far more powerful and direct evidence for the corpuscular hypothesis than any amount of logical deduction from first principles (Shapin and Schaffer 1985, 40, 169–185).

Neither the experimentalists nor their opponents used the language of aesthetics as philosophers understand it today. It was not until the 18th century that European thinkers began to carve out a special field of inquiry devoted to the affective states arising from our sensory encounters with external things. Nevertheless, disagreements about whether such experiences were helpful or harmful were of great importance to the debates about experimental science that took place during the 17th century. After all, the key difference between experimentalism and the position held by thinkers like Descartes (later termed rationalism by intellectual historians) was that the former set a high value on the evidence of the senses, while the latter regarded that evidence as not only unreliable but also disruptive to the more trustworthy intellectual efforts needed to discover the truth. For Descartes, discovering the truth depended in large measure on disaggregating the pleasures and pains of the senses from the mental processes of deduction – processes that he regarded as belonging to the immaterial soul, and thus having a life beyond that of the body. Meanwhile, experimentalists like Boyle proposed a form of inquiry that contemporaries regarded as sensual, leaving the philosopher exposed and perhaps even subject to all the pleasures and pains that sensory experience might provoke. After all, one widespread criticism of the Royal Society, frequently declaimed by hostile churchmen, held that experimentalism was nothing more than a species of epicureanism – a voluptuous preoccupation with the corporeal pleasures of the senses that served only to distract its proponents from the worthier contemplation of an incorporeal God. Such criticisms were made all the more convincing by the fact that Boyle's corpuscular theory of matter resembled that of the ancient Greek atheist Epicurus. Epicurus hypothesised a world brought into being between random collisions of imperceptible atoms in order to urge that the gods have no control over the world and that we should therefore feel free to enjoy the pleasures of the senses without fear of divine retribution (Clericuzio 1990, 571–573; Wilson 2008, 237–240).

It is true that members of the Society recognised the threat posed by these accusations, and that they tried to portray the sensory pleasures of experimental science as innocent forms of gratification that would ultimately bring the natural philosopher closer to God. But at the same time, some of them packed their writings with vivid evocations of those pleasures, often blurring the lines between the supposedly disembodied delights of the mind and the bodily gratifications of the senses. As Elizabeth Swann (2020) has recently pointed out, Robert Hooke began his *Micrographia* (1665) with the startling image of the natural philosopher reversing the effects brought about when Adam and Eve ate the forbidden fruit by not only contemplating but also eating natural phenomena:

> And as at first, mankind *fell* by *tasting* of the forbidden Tree of Knowledge, so we, their Posterity, may be in part *restor'd* by the same way, not only by *beholding* and *contemplating*, but by *tasting* too those fruits of Natural knowledge, that were never yet forbidden
> (Swann 2020, 144. See Hooke 1665, sig. [b2r-b2v])

Boyle did not trade in such daring images as Hooke, but he nevertheless repeatedly coached his readers to enjoy experimental science. In *The Origine of Formes and Qualities* (1666), the work we'll dwell upon most here, he endlessly recapitulated the delight he had taken in the phenomena of chemical experimentation. Consider his remarks on the chemical production of blue vitriol – a substance we would call copper sulphate:

> yet repeating the Experiment with care and watchfulness, we, this way, obtain'd one of the loveliest Vitriols that hath perhaps been seen, and of which you your self may be the judg by a parcel of it I keep by me for a Rarity.

As he did elsewhere, Boyle here insisted that chemical experimentation would lead to sensory pleasure. And note his insistence on not only the loveliness of the vitriol but also its rarity. There is a hint here at a kind of beauty that might attract monetary value. Boyle's account places us firmly in a sensual world of pleasure taken in material things, a long way from disembodied acts of pure intellection (Boyle 1666, 217–218).

Since the 1980s, historians and sociologists of science have put forward contrasting explanations for this interest in the pleasures of the senses. For Lorraine Daston and Katharine Park (2001), it exemplifies a tendency among scientists of the 17th century to regard the wonder stoked by rare and marvellous objects as a necessary stimulus for the model of scientific inquiry that they promoted. Boyle thus portrayed the apparently ignoble objects to which he devoted so much attention – such as rotting fish – as worthy of attention because of the unexpected pleasure that they could provoke if skilfully observed (Daston and Park 2001, 311–312). By contrast, Steven Shapin (1984) and Lawrence M. Principe (1995) have addressed the matter through Boyle's style of writing, showing that he used rhetorical devices then widely thought to reproduce the same pleasures of sensory experience. The similarities end here, however. Shapin regards Boyle's vividly sensorial descriptions, rich in circumstantial detail, as a technology of "virtual witnessing". On this view, Boyle saw his style as a means of making readers feel as though they could have witnessed the experimental phenomena for themselves, thereby winning them over to the experimental form of life (Shapin 1984, 490–497). But Principe disagrees with this assessment, preferring instead a more prosaic explanation. He portrays Boyle's literary style as the by-product of his encounter with French heroic romances when he was young, during the 1630s and 1640s. On this view, far from seeking to compel assent through a literary technology, Boyle simply adopted a readable style then popular among people of his social class and education (Principe 1995, 395–397).

This chapter will reconcile these apparently contrasting positions, showing that Boyle and his contemporaries closely linked the pleasures of rhetoric to the production of knowledge about natural things. This argument will consist of two parts. In the first, we will see how Boyle used experimental descriptions to produce vivid mental images, both of the phenomena he had actually witnessed, and of the imperceptible processes that he thought had caused them. Here, we will go beyond the now relatively familiar observation that many thinkers of the 17th century saw putatively vivid descriptions as tools for an overpowering reason and compelling assent (Gaukroger 1992, 591–596; Wragge-Morley 2020, 126–130). Rather, we will see that Boyle regarded such descriptions as means of demonstrating that natural phenomena were caused by the agents he hypothesised – the substrate of imperceptible particles that he called corpuscles. In the second part, we will see that Boyle also enlisted the experience of beauty itself to argue for the significance of his experimental findings. For Boyle, beauty could enable the experimentalist to distinguish between the products of art and those of nature. And in turn, the ability to make this distinction could enable the philosopher to use chemical experiments as arguments for the corpuscular theory of matter. Boyle, therefore, set a high value on the cognitive effects of sensory pleasure. As we'll see, he held that pleasurable forms of experience could enable the philosopher to understand what kinds of agents had brought the phenomena of experiment into being. As a result, he suggested that such pleasures could help the philosopher reckon with fundamental questions about the causes of change and continuity in natural things.

3.2 What Makes a Substance? The Contemporary Debate

To understand why Boyle attached such value to sensory pleasure, we must first understand what he sought to prove – and disprove – with his programme of chemical experimentation. Robert Boyle described the experiments I'll talk about today in a book called *The Origine of Formes and Qualities*, published in 1666 – the same year as the Great Fire of London. Here, Boyle sought to do two things. First, he wanted to discredit a framework inherited from the medieval and early modern interpreters of the ancient Greek philosopher Aristotle for explaining what makes substances the way they are. That framework is known as hylomorphism, and it supposes that natural things are made what they are by an immaterial form. It is the immaterial form that makes sense of matter and gives rise to distinctly different kinds of substances, such as mercury or gold. Second, he wanted to advance a replacement explanation for why substances are the way they are. This explanation is known as the corpuscular hypothesis because it holds that the things we perceive as substances get their substance-ness from the interactions between imperceptibly small particles of matter – particles that Boyle called corpuscles. Before turning to the experiments with which Boyle evidenced his position, let me briefly explain how it differed from the hylomorphism he opposed (Garber 2013, 6–7).

In *Formes and Qualities*, Boyle portrayed the hylomorphic position as one that had changed little since the crucial interventions of medieval scholastic thinkers such as Thomas Aquinas. For Aquinas, matter on its own was an impossibility, having no form whatsoever. He could only hypothesise its existence as the potential to become all kinds of different things. Like Aristotle, however, he held that the natural world is made up of substances that are essentially different from each other and that those entities have a tendency to purposive activity. For prime matter to become anything at all, it, therefore, required what Aristotle called a substantial form – an agent that somehow gave matter identity and intentionality (Newman 2006, 36–38, 107–108). Aquinas thus famously argued in his *Summa Theologica* that the immortal soul was the substantial form of the human being. Without that soul or substantial form, the human body would be something else entirely – or nothing at all (Aquinas 1920–1922, vol. 4, pp. 3–54). Aquinas and many of his later followers thought about substances such as gold or mercury in pretty much the same way. Those things got their essence from the substantial form, which mysteriously differentiated them from the rest of the matter in the universe, and gave them the qualities for which we know them. And scholastic thinkers were careful to separate out the qualities attributed to the substantial form from those that arose by accident, or as the result of human intervention. As they saw it, only natural things had substantial forms. Consequently, they regarded artificial things as categorically distinct from the works of nature (Newman 2006, 107–108, 114–115).

By contrast, Boyle held that substances get their properties from the composition of, and interactions between, imperceptibly small particles of matter called corpuscles. For our purposes, we need to understand two key aspects of this hypothesis. The first was that Boyle regarded the corpuscles not as the smallest, most fundamental particles of matter. Rather, he regarded them as larger semi-permanent entities arranged out of those particles to make up individual substances such as gold and mercury (Anstey 2000, 44, 63; Newman 2006, 179–180). Second, he regarded the transformations that we can perceive in natural things as being caused by the mechanical interactions between those particles. Consider for instance what happens when gold is dissolved in a mixture of nitric acid and hydrochloric acid, a mixture known as *aqua regia*. According to Boyle's caricature of the doctrine of forms, you would need to suppose that dropping the gold into the aqua regia somehow annihilated the substantial form of the gold, or at least prevented it from operating. How otherwise could you explain the gold's disappearance? But Boyle held that the same phenomenon could be explained by appealing to changes in the texture of the corpuscles of gold and aqua regia that were brought about when those corpuscles collided with each other. He therefore rejected the idea that substances came into being through the agency of immaterial and in some sense intelligent entities. Instead, he preferred to think about substances as if they behaved in the same way as physical things accessible to sensory experience. He thus talked about those imperceptible particles of matter in physical terms, hypothesising that substances depended on differences in their motion, composition, and resulting textures (Boyle 1666 189–190; Newman 2006, 174).

Given that Boyle, like other philosophers of his stamp, regarded the corpuscles as imperceptible, it is a paradox that he thought it possible to learn about them through sensory experience. But we can begin to appreciate his position by returning to his attack on hylomorphism. For Boyle, the problem with substantial forms was that they were unintelligible. Indeed, he saw the doctrine of forms as a kind of non-explanation, shedding no additional light on the processes involved. By contrast, he frequently stated that the advantage of the corpuscular hypothesis was not so much its truth value, but rather its superior intelligibility (Wragge-Morley 2020, 40–41). This was in fact a position that Boyle shared with those modern philosophers who supported versions of the corpuscular hypothesis while denying that experiment could reveal anything of importance about it – figures such as Descartes and Thomas Hobbes. Despite disagreeing about the value of experiment, they shared the position that explanations for physical phenomena must be couched in physical terms and that therefore, in practice, they had to be derived from phenomena accessible to sensory experience. Consider, for instance, Descartes's strategy in *Le Monde*, first published posthumously in 1664 but composed from 1629–1633. There, he argued that we can account for all the physical phenomena by hypothesising a universe made up of innumerable tiny particles of matter and imagining how those particles of matter would behave if set into motion. He could reduce those motions to a simple set of rules – or natural laws – derived from our regular sensory experience of bodies in motion. We can figure out, for instance, that bodies have a tendency to move in a straight line unless somehow interrupted by another body when we think of what happens when a stone is projected from a slingshot. As soon as the stone is released, the circular motion caused by the constraint of the sling is turned into a linear motion as the stone is shot into the air (Descartes 1998, 29; Dear 2008, 20–24; Adams 2014, 411, 419–420).

For its intelligibility, therefore, the corpuscular hypothesis depended in a very specific manner on the senses. The explanations that it offered got their persuasive force from our apparent tendency to conceive physical causation in terms derived from past sensory experience – along with our apparent inability to imagine them in any other terms. Where Boyle differed from Descartes and Hobbes, however, was in his assertion that phenomena accessible to the senses could also furnish reliable evidence for that hypothesis (Wragge-Morley 2020, 50). In other words, he claimed that phenomena accessible to the senses could show not only that imperceptible corpuscles actually existed, but also that those corpuscles operated according to the apparently familiar principles of mechanics. For Boyle, there were therefore two senses in which sensory experience was important. First, sensory experience gave the corpuscular hypothesis its explanatory force, providing the mind with the tools it needed to come up with plausible mechanisms of physical causation. Second, it somehow gave the mind an embodied connection to imperceptible particles of matter, enabling it to reckon with them through the more familiar objects of sensory experience. As we will see, recognisably aesthetic forms of experience were at stake in this regime of perception and cognition, with sensory pleasure playing an important role in making imperceptible corpuscles into objects of knowledge.

3.3 Pleasure, Pain, and the Identification of Substances

In *The Origine of Formes and Qualities*, Boyle made it clear from the outset that he intended to mobilise particularly vivid and pleasurable forms of sensory experience. Consider, for instance, how he ended the book's preface, addressed to a fictional interlocutor named *Pyrophilus* (lover of fire):

> And though I know, that some of the things my Experiments tend to manifest, may likewise by confirm'd by the more obvious Phænomena of Nature, yet I præsume you will not dislike my chosing to entertaine you with the Former […] because the changes of Qualities made by Our Experiments will for the most part be more Quick and Conspicuous, and the agents made use of to produce them, being of our own Applying, and oftentimes of our own Preparation, we may be thereby assisted the better judge of what they Are, and to make an æstimate of what 'tis they Do.

Here, Boyle noted that the controlled circumstances of an experiment made it easier to identify the exact agents involved in producing a given effect. And he noted such effects were frequently far more palpable – 'more Quick and Conspicuous' – than the ones we encounter in everyday life

(Boyle 1666, sig. [B7 r-v]). Like the more recent experimenters discussed by Milena Ivanova in this volume and elsewhere, Boyle designed experiments to be clear and vivid in their effects, hoping to convince observers and readers through rapid and obvious results (Ivanova 2021, 4–5).

We can get a sense of what Boyle had in mind by turning to the seventh of the experiments that he appended to the text in order to illustrate his position. Boyle and his assistants began the experiment by preparing a *menstruum peracutum* – a powerful solvent – by distilling a liquor out of a mixture of butter of antimony (antimony trichloride) and spirit of nitre (ethyl nitrate). Next, they took small morsels of highly refined gold, taking care to eliminate any impurities, and put them in the menstruum. At this point, they sometimes added a little spirit of salt (hydrochloric acid) to get the reaction going. Although the gold dissolved in this mixture, they nevertheless found at the bottom of the vessel a white powder upon which the menstruum seemed to have no effect. Using borax as a flux, they then melted down the powder into a metal. On inspection, they found that it had all the sensory qualities that one would expect to find in silver. It was silvery-white, they could flatten it with a hammer, and, when they dissolved it in *aqua fortis* (nitric acid) or spirit of nitre, the metal produced an "odious bitterness". In other words, Boyle claimed that he had successfully turned gold into silver, effecting a transmutation somewhat resembling the ones that had long been sought after by generations of alchemists (Boyle 1666, 349–378. On the qualities of the silver, p. 354).

As Lawrence M. Principe (2000) and William R. Newman (2006) have both shown, Boyle devoted much effort to the alchemist's goal of turning base metals into gold – a goal that was in any case consistent with his theory of matter (Principe 2000; Newman 2006, 157–189). Here Boyle wryly noted, however, that there wasn't much money to be gained from this particular transmutation, in which he appeared only to have made gold into a less valuable metal. But he nevertheless regarded the experiment as a powerful demonstration that the doctrine of substantial forms lacked explanatory value (Boyle 1666, 367). Here, the choice of gold was significant. Gold is of course valued for its tendency to remain the same, and Boyle thus parodied the scholastic position by stating that 'there is not any [substance] whose Form is more strictly united to its Matter'. If you held the scholastic position, in other words, you would suppose that the form of gold was particularly strong, giving the substance its unusual resistance to destruction. Yet Boyle had used chemical agents to turn gold into another substance entirely, overcoming the apparent tenacity of the substantial form. At the same time, he had used those chemical agents to make a new substance entirely – silver. He, therefore, reasoned that you don't need to hypothesise the activity of substantial forms to explain what gives gold its gold-ness or silver its silver-ness. Instead, he framed an explanation based on the interventions through which he had himself brought about the destruction of the former, and the surprising appearance of the latter (Boyle 1666, 364).

Before considering that explanation, we must first figure out how Boyle attempted to demonstrate that he had truly turned gold into silver. It was on this point, after all, that Boyle's argument depended. He pointed out that the metal in question was white, malleable, dissoluble in aqua fortis and spirit of nitre, and bitter upon that dissolution. In other words, he cycled through several different forms of sensory experience, including vision, touch, and taste or smell (Boyle 1666, 354). Here, Boyle deployed an approach to the definition of substances that would be taken up by John Locke in his famous expression of the empirical position, the *Essay Concerning Human Understanding*. On this view, there was no single essential property – such as a substantial form or single physical feature – that gave an entity its specific identity. Instead, Boyle and Locke adopted a nominalist position, assuming that the names or other signifiers that we give to substances reflect our experiences of them rather than some essential feature. As a result, they held that identifying substances by a single form of sensorial evaluation carried too much risk of misidentification. There was simply no way to be sure that a single form of sensory experience would yield reliable insights into the essential properties of the object in question – those properties that truly distinguished the object from any other. Whether understood either as a substantial form or as the composition or texture of imperceptibly small particles of matter, such essential properties were thought wholly inaccessible to the senses. The solution used by Boyle and Locke, although of course not devised by them, was to enumerate many more of the properties through which the senses enabled the mind to distinguish one substance from another. The idea was that relying on a wider range of such properties would reduce the risk of mistakenly conflating superficially similar substances with each other (Yolton 1970, 42; Anstey 2011, 51–59; Wragge-Morley 2020, 123–124).

But note that Boyle's enumeration of the silver's properties did not pretend to evoke some objective form of sensory experience. It was also decidedly sensuous. Although it began with vision, a sense then widely considered somewhat comparable to the disembodied intellect, it also included three senses then widely associated with the body's animal desires for food and sex – touch, taste, and smell (Von Hoffman 2016, 73–79). Moreover, Boyle remarked not only on the bitterness of the silver but also on the "odious" unpleasantness of his encounter with it. He filled *Formes and Qualities* with many more instances of sensory pleasure and displeasure, confirming our impression that such apparently subjective forms of judgement had a role in his form of empiricism. Throughout the text, his descriptions of the apparent bitterness, sourness, and sweetness of substances rapidly shaded into judgement so of their pleasantness and unpleasantness. In addition to describing chemical compounds as 'strong' or 'strongly' smelling or tasting, he thus added qualifiers such as 'piercing', 'offensive', or 'pungent' (Boyle 1666, 139, 273, 277, 280, 309, 314, 315, 324, 340, 427, 428, 430). And he was by no means shy of describing how his body, and those of other onlookers, responded to those sensations. In the tenth experiment, he thus gave a lively account of the liquor that emerged as the final by-product of distilling spirit of wine (an aqueous solution of ethanol) together with oil of vitriol (concentrated sulphuric acid):

> after this volatile and odoriferous Spirit is come over, and has been followed by an Acid Spirit, it will usually, towards the latter end of the Distillation, be succeeded by a Liquor, that is not onely not fragrant, but stinks so strongly of Brimstone, that I have sometimes known it almost take away the Breath (as they speak) of those, who, when I had the Receiver, newly taken off, in my hand, did (either because to make sport I gave them no warning, or because they would not take it, as thinking what I told them impossible,) too boldly adventure their Noses in the Tryal.

In this account, Boyle framed the body as a kind of scientific instrument, registering its response to the liquor's potent, sulphurous aroma. But this body was hardly objective, at least not in the modern sense of the term. It was an affective instrument, distinguishing one substance from another through its embodied judgements of pleasure and pain, along with other non-cognitive responses – here, a near loss of breath – to the objects of experience (Boyle 1666, 428–429).

To a large degree, this model of embodied empiricism lines up with what historians including Lorraine Daston, Peter Galison, Lissa Roberts, and others have told us about the pre-modern empirical sciences of the 17th and 18th centuries. Daston and Galison (2007) have thus shown that the empiricism of the 18th century depended not on the kind of objectivity that would later be associated with the use of precise measuring instruments, but instead on the exercise of judgement by natural philosophers about the meanings of their experiences, and those of other observers (Daston and Galison 2007, 66, 74). Meanwhile, Roberts (1995) has shown that, despite their attempts to portray their work as cerebral and disembodied, the exercise of judgement was frequently a matter of calibrating the body's sensuous encounters with the world. Indeed, she has shown that such a sensuous mode of inquiry was the dominant form of empiricism employed by chemists until the latter part of 18th century, with individuals using their whole bodies to make fine qualitative distinctions between substances. It was not until the 1770s that figures such as Antoine Lavoisier (1743–1794) began to practice a recognisably modern form of chemical empiricism, based instead on the quantitative data yielded by laboratory instruments (Roberts 1995.

See also Giglioni 2006). If Boyle's empiricism was thus recognisably pre-modern, it was also perhaps distinctive for its insistence on the epistemic value of judgements that most modern readers would probably identify as belonging to the fields of taste or aesthetics. The pleasures and pains of sensory experience were not merely adjuncts to the exercise of embodied empiricism, occasionally noted in passing. Instead, such judgements helped him to distinguish one substance from another. In the eighth experiment, to take just one more example, he used the "annoyance" provoked in the tongue and nose by a substance to help prove that it was aqua fortis. For Boyle, the body's experiences of pleasure and pain were tools for scientific discovery, helping him and his readers vividly distinguish one substance from another – and thus to better judge the outcomes of chemical experiments (see Wilson 2008, 230–246).

3.4 The Embodied Experience of Mechanical Causation

As we have seen, Boyle wanted his experiments to do far more philosophical work than to show that certain reactions had taken place and that certain substances had been produced. He wanted them to serve as evidence that those changes had been brought out by mechanical interactions between the imperceptibly small corpuscles out of which he believed substances to be composed (Newman 2006, 3–4, 175–189). To some modern-day readers, it might come as a surprise to find that concerns about the pleasures of the imagination, articulated through theories of rhetoric, were important to debates about corpuscular and atomic theories of matter. But this connection may have seemed more obvious to thinkers of the 17th century, who all knew that one of the key ancient sources for the corpuscular hypothesis was the Roman poet Lucretius's *De Rerum Natura* ("On the Nature of Things"), a philosophical poem renowned as much for its beautiful and striking imagery as for its moral, medical, and scientific insights. And of course they also knew that the ancient Greek thinker Epicurus had proposed this version of atomism to demonstrate that the gods have no role in the affairs of the world, and thus to free people from fear of divine retribution if they indulged in the pleasures of the senses (see Gillespie 2007 and Hock 2021; on attempts to circumvent the charge of hedonism labelled at those who followed Epicurus's matter theory, see Garau 2017, 240). In addition, there were obvious albeit superficial parallels between the method that Lucretius had used to evidence his hypothesis, and the one employed by Descartes. The two authors both employed a strategy that was basically conjectural, putting forward an account of matter and motion-based not on systematic observation but rather on its capacity to account for how natural phenomena appear to us and generally persist in that state. Lucretius accomplished this end by positioning his theory of matter as poetic fancy, albeit with a claim to the truth. Meanwhile, Descartes positioned his theory as a hypothesis – an account based not on historical truth but rather a conjecture based on what the intellect can figure out about the properties of matter and the laws governing its motion. In turn, this intellective act enabled him to explain precisely how mechanical interactions between particles of such matter might give rise to the world as we know it (Descartes 1998, 21; Wilson 2007, 136–138).

In effect, this method enabled Descartes to suppose that mechanisms analogous to the ones we know from experience were in operation among the fundamental-but-imperceptible particles upon which hypothesis depended. He famously insisted, for instance, that magnetism was caused by invisibly small helical particles that pass through threaded pores in magnetically charged bodies. He thus held that the particles were pushed out of the body at one pole, and pulled in at the other, mechanically producing the otherwise mysterious forces of attraction and repulsion we associate with magnetism. But many of Descartes's contemporaries worried that these hypotheses were little more than figments of his imagination. Matthew L. Jones reminds us, for instance, that the Dutch university professor Martinus Schoock (1614–1669) took Descartes to task for putting comparisons in the place of philosophical reasoning (Jones 2006, 74–75). For Schoock, the problem was that these comparisons were nothing more than rhetorical tools, wholly inappropriate to the task of rational demonstration:

> Comparisons, serving above all to illustrate certain statements, are instruments for the orator who seeks to use appropriate words for persuasion, rather than for the philosopher who pursues the essence of things that he seeks to prove through solid reasons. Also the comparison [...] neither proves nor confirms what is in question, at most illustrating the dogma itself, or elucidating that which has been demonstrated
>
> (Descartes and Schoock 1998, 253)

There was nothing novel or unusual about Schoock's definition of the comparison. In works of rhetorical theory, both those written at the time and their ancient antecedents, the comparison was defined as a figure of speech intended to make something difficult easier to understand. Its purpose was to relax the mind, using the image of something familiar to make it easier to picture something unknown or difficult to grasp. By characterising Descartes's explanations as mere comparisons, he, therefore, sought to dismiss them as distracting appeals to the pleasure-seeking imagination rather than the more discerning intellect (Walmsley 2003, 108–117; see also Galison 1984, 323; Wragge-Morley 2020, 132–143, 150–152).

For the most part, Boyle and his contemporaries bought into this criticism, characterising Descartes's explanations as ingenious but ultimately unfounded hypotheses (Sargent 1995, 35–37). But Boyle asserted, as we have seen, that he had experimental evidence for his own version of the corpuscular hypothesis. The problem, however, was that this evidence was at best indirect. His experiments simply did not make the imperceptible corpuscles he hypothesised accessible to the senses, and he was too acutely aware of the limits and nature of sensory experience to think that he could gain direct access to them. All Boyle had, in other words, were phenomena from which the existence, properties, and motions of those particles might be inferred. One way of responding to this predicament was through modesty. Unlike Descartes, Boyle rarely framed specific micro-mechanical explanations for experimental phenomena, for the most part preferring only to argue that their causes were in some sense mechanical. But even making this argument involved, Boyle in the kind of analogical reasoning that might, in the terms of 17th-century rhetorical theory, shade into figurative speech (Boyle 1666, 366; Shapin and Schaffer 1985, 50–55).

Let's return to the alleged transmutation of gold into silver. After describing his materials, procedure, and the phenomena yielded by the experiment, Boyle started to work out its philosophical significance. Perhaps the most important was that even the forms of tenacious substances such as gold could be altered by apparently mechanical means:

> at least among inanimate Bodies, the noblest and constantest sort of Forms are but peculiar Contrivances of the Matter, and may, by Agents, that work but Mechanically, that is, by locally moving the parts, and changing their Sizes, Shape, or Texture, be generated and destroy'd; since we see, that in the same parcel of Metalline Matter, which a little before was true and pure Gold, by having some few of its parts withdrawn, and the rest transpos'd, or otherwise alter'd in their structure [...] or by both these ways together, the Form of Gold, or that peculiar Modification which made it Yellow, indissoluble in *Aqua fortis*, &c. is abolished, and from the new Texture of the same Matter, there arises that new Forme, or Convention of Accidents, from which we call a Metal Silver;

In this passage, Boyle slips between his account of his own procedures and his hypothesis about what has taken place at the corpuscular level. He characterises his own experimental procedure – dissolving gold in the *menstruum peracutum* – as a series of mechanical interventions, consisting in moving the substances around and thereby changing their sizes, shapes, and textures. Immediately, he asserts that something similar has gone on among the imperceptible corpuscles whose existence and attributes he could only guess at. Some of the corpuscles of gold had also been moved around or otherwise altered, resulting in a substance with a different structure – silver (Boyle 1666, 365–366).

You might say that Boyle had more warrant than Descartes for his hypothesis. Boyle could at least plausibly argue that the apparent transformation of gold into silver had happened after, and therefore likely because of his mechanical interventions. That said, the experiment does not provide any direct evidence that infinitesimally small corpuscles behave in the same mechanical fashion as Boyle's sensorially accessible interventions. Instead, it invites us to frame an analogy between the world of imperceptible entities and the world of objects that we can apparently perceive and understand. It then seems fair to say that Boyle's chemical experiments worked in part by presenting onlookers or readers with sensorially accessible models for visualising, and therefore understanding processes that would otherwise have remained wholly beyond the reach of the senses. Elsewhere, I have shown that Boyle understood how the use of such analogical and comparative reasoning strategies left him open to precisely the same kind of criticism that Schoock had earlier levelled at Descartes – namely, that he had substituted pleasant images for real explanations. Strikingly, however, Boyle responded by stating that such explanations were valuable even if they could not be proven true. His point was that coming up with a plausible mechanism for a given phenomenon established that mechanism as a possibility. If the philosopher could figure out how a mechanism might give rise to the phenomenon in question, then she or he had established both that the mechanical model had explanatory power, and that it might conceivably be true (Mulligan 1994, 254–255; Wragge-Morley 2020, 71). In a sense, therefore, he sought to exploit the elucidatory power for which the comparison was recognised in contemporary rhetorical theory. By finding analogues for the interactions between corpuscles in phenomena accessible to the body's sensory organs, Boyle gave the mind easy access to his mechanical model of physical causation. But at the same time, like Descartes, he apparently regarded the framing of such analogues as philosophically productive in its own right, establishing the very possibility that nature could be mechanical. To be capable of coming up with mechanical explanations for natural processes was a way of establishing that nature was itself mechanical.

3.5 Beauty and the Naturalness of Phenomena

So far, we have seen that Boyle used the affective dimensions of sensory and imaginative experience to make two distinct kinds of knowledge claims. He used the pleasantness and unpleasantness of experimental phenomena, along with a host of other apparently subjective experiences, to make judgements about the identity of chemical compounds. And, as we have just seen, he used the affective power of the comparison to make suggestive claims about the ontology of nature, with mechanical hypotheses about natural processes serving to vindicate the idea that nature might itself be mechanical. We have not yet, however, encountered aesthetic experience in the sense with which most readers today would be familiar – judgements of beauty and ugliness. But from time to time, Boyle also used the experience of beauty to advance a scientific argument about the results of his experiments, and the value of those results their value as explanations for the nature of substances.

From the Aristotelian perspective, the obvious response to Boyle's mechanical explanation for substances came from the distinction between nature and art – a distinction that we have already encountered. This distinction is more or less irrelevant to the sciences today, but it was important to Aristotle's manner of explaining natural phenomena. According to Aristotle, natural things had a fundamentally developmental quality, tending towards the realisation of some kind of potential. Consider the example of a tree. For Aristotle, the seed of that tree was endowed with a principle of life enabling it to realise the potential of becoming a tree. The matter of the tree was not stable over time (think of the huge difference between a seed and a whole tree), but it somehow possessed a principle giving it both continuity of identity and the ability to realise its potential. As we have already seen, many scholastic thinkers argued that the substantial form was this principle. In addition, they tended to see substances such as metals and minerals in the same animistic terms – even though we would today regard them as inorganic, and thus not living. By contrast, Aristotle saw artificial things as objects lacking this principle of life. Consider a table. Given its shape by human agency, that table has no innate tendency towards the realisation of any potential, and it will always remain static. The scholastics thus saw artificial things as in some sense dead, lacking the principle of life that gave natural things their developmental character (Aristotle 1996, II.1, 33–35).

Boyle realised that his scholastic opponents might therefore object that the substances obtained from his experiments – such as the silver he obtained from gold – were artificial, and therefore irrelevant to explaining how natural things worked. The scholastics might very well point out that they fully expected the artificial productions resulting from experiments to lack substantial forms, and therefore discount them as evidence against their hypothesis. Now, it is well known that Boyle and other followers of the new philosophy weakened this distinction considerably – the mechanical model depended instead on regarding most of the causes of artificial and natural things as ontologically identical (Dear 1995, 151–161). In *Formes and Qualities*, therefore, Boyle went to great lengths to demonstrate that his substances were just as natural as those that might be encountered outside the laboratory. Indeed, we have already encountered Boyle's principal strategy for making this argument – challenging his readers to distinguish the artificial substance from the natural one. This was exactly what he did with the silver, enumerating the qualities of his artificial silver to show that no person would be able to distinguish it from any other piece of silver. To all intents and purposes – at least insofar as the senses could discern – it was natural silver.

At other times, however, he asserted that his substances must be natural because they were just as elegant and beautiful as their natural counterparts. This was a point that Boyle made, for instance, in a section of the work dealing with the regularity and beauty of the crystals of salts and metals – characteristics that some philosophers had used as evidence for the activity of the substantial form or some other guiding power. Among the examples he cited was a well-known experiment that involved suspending a thin slice of copper in a solution of silver and aqua fortis (nitric acid). Soon, tiny plates of silver attached themselves to the copper:

> [...] there would settle, all about it [the copper], swarms of little Metalline and Undiaphanous Bodies [of silver], shining in the water like the scales of small Fishes, but form'd into little Plates extremely thin, with surfaces not onely flat, but exceeding glossy: and among those, divers of the larger were prettily figur'd at the Edges.

This passage relies on the rhetorical force of the comparison. Boyle uses something familiar – the scales of fishes – to help the reader more easily form a mental image of the slender, smooth, shiny, and geometrically precise silver crystals that he had encountered (Boyle 1666, 231). The rest of this section is packed with similar such evocations, each insisting that experimentally produced crystals are just as 'fine and exquisite', 'smooth & specular', or even just as 'lovely' as those to be found out in nature (Boyle 1666, 230–240).

For Boyle, the loveliness of these crystals served his broader argument about the causes of change and continuity in substances. He asserted that the scholastics had seen the surpassing beauty of crystals as evidence that the intelligent agency of substantial forms must be in play. There was no way, they reasoned, that unintelligent material causes could give rise to such symmetry and order. But Boyle had apparently succeeded in producing crystals that were just as pleasing to the senses:

> [...] If then these Curious shapes, which are believed to be of the admirabilest Effects, and of the strongest Proofs of substantial Forms, may be the Results of Texture, and if Art can produce Vitriol itself, as well as Nature; why may we not think, that in ordinary *Phænomena*, that have much less of wonder, recourse is wont to be had to substantial Forms without any Necessity?

Once again, Boyle mobilised aesthetic experience in the service of a scientific argument. He had used mechanical agents – this is what he means by "Texture" – to produce crystals that were indistinguishable from their naturally occurring counterparts. They exhibited exactly the same degree of beauty and perfection. It stood to reason, therefore, that philosophers did not need to hypothesise mysterious substantial forms to explain how such bodies came into being. Instead, they could simply refer to the mechanical processes through which Boyle had made them (Boyle 1666, 241). As we have seen, asserting the likeness between experimental procedures accessible to the senses and the workings of imperceptible particles that nobody had ever seen, with contemporaries noticing that the reasoning was analogical rather than empirical. Nevertheless, Boyle hoped that the beauty of his crystals would make that analogy more plausible. Using beauty as an outward sign of inward similarity, he hoped to convince his readers that the mechanical means he had used to make his crystals were like the processes through which they naturally came into being.

3.6 Conclusion: Aesthetic Experience and the Ontology of Nature

We can now see that Boyle's endless evocations of the beauty and pleasure to be had from experimentation served philosophical purposes. As Shapin has shown, Boyle used descriptions rich in sensorial detail to give readers the impression that they were witnessing the experiment for themselves. What is more, the use of vivid comparisons between phenomena accessible to the senses and the processes going on among imperceptible corpuscles had an important cognitive function, helping readers to more easily conceive of explanations framed in mechanical terms. I am not aware, however, that anybody has discussed the way Boyle's explanatory strategies – themselves highly rhetorical – depended on the apparently subjective forms of experience arising from the likes and dislikes of the senses, or the encounter with beauty. In other words, it is fair to say that Boyle conceived of his chemical experiments in terms that we would recognise as belonging to the domain of aesthetics. In its most important aspects, however, Boyle's engagement with aesthetic experience has a decidedly pre-modern quality. This was because he did not insist on the modern distinction between the subjectivity of universal judgements and the objectivity to be gained from scientific investigation. Rather, he sometimes used those experiences to advance scientific arguments about the identity of substances, as well as their physical causes. In other words, Boyle regarded the sensuality of scientific experience – something for which, as we have seen, he was widely criticised – as an important part of the cognitive work involved in making scientific judgements.

Boyle was by no means the only natural philosopher of the 17th century to see aesthetic experience as a tool for making judgements about natural phenomena. As I have shown elsewhere, Boyle's immediate contemporary and colleague Robert Hooke made similar claims for beauty in his *Micrographia*, first published in 1665. In that beautifully illustrated study of microscopically small phenomena, Hooke used a geometrical standard of beauty to distinguish between the works of art and those of nature. He thus famously opened the book by revealing that some of the most highly refined objects made by humans – a razor, the point of a needle, and printed full-stop – were in fact ugly under magnification. By contrast, even the least promising works of nature, such as fleas, lice, and flies, turned out to be wonderfully beautiful. What is more, saline crystals had an important role in Hooke's argument. He identified their structures as evidence of God's designing agency, their beautiful and precise geometrical forms expressing a mathematical order upon which everything in nature seemed to depend (Hunter 2010; Wragge-Morley 2020, 7–8, 84–90). This argument in fact contrasted in one sense with the one Boyle put forward. Where Hooke used the experience of beauty to distinguish art from nature, Boyle made the opposite move, bringing the two closer together. The difference between these two positions, however, reflects a tension identifiable in the works of both philosophers. On the one hand, they both argued for a position that regarded most of the causes of artificial and natural phenomena as belonging to the same ontological category. Yet they both also regarded God as a supremely talented and powerful creative agent, far outclassing humans in his ability to design and make things. To make the case that the causes of natural and artificial things were alike, Boyle had to assert that substances he had produced were nonetheless natural. He had to insist that they exhibited the same beauty and order to be found in every other product of God's creative activity.

This belief in God's creative involvement with the world enabled Hooke and Boyle to regard aesthetic judgement as a means for making scientific arguments. In some respects, after all, their position on the meaning of sensory experience was relativistic. They understood that perception arose from the interactions between external things and the body and that in consequence sensory experience might vary greatly between species and between individuals. They knew, for instance, that properties such as redness did not belong to the objects we denominate as red, but instead arise in us as a response to some external things. It was for this reason that they rejected the essentialism of substantial forms, preferring instead to see substances as names given to different combinations of sensory experience, probably – but not necessarily – corresponding to the essences of the things themselves (Sloan 1972). At the same time, however, believing that the human sensorium was part of the same creation designed by a supremely powerful, wise, and benevolent God made it possible for Boyle and Hooke to see more continuity between aesthetic experience and its objects than might be admitted today (Brooke 1991, 131–135). It was possible, in other words, for these two philosophers to see the experience of beauty as a divinely-instituted symptom of the encounter with natural things, enabling the discriminating observer to make accurate judgements about their mode of being. In turn, this appeal to a fundamental order underlying creation may have made it easier to present some forms of sensory pleasure as philosophically and socially legitimate, betokening not the wonton sensuality associated with Epicureanism, but instead an innocent delight in God's creation.

It would be mistaken, however, to assume that Boyle's engagement with aesthetic experience has no lessons for people today. It is true that few would now accept the metaphysical framework that made it possible for him to regard the experience of beauty as sufficiently fixed for it to form the basis for scientific knowledge. But his seeming recognition that we cannot disaggregate the cognition of external things from their effects on our minds and bodies remains important. Indeed, it resonates with recent developments in neuroscience and philosophy of mind, with thinkers such as Alva Noë and Mark Johnson increasingly advocating a model of the mind that includes not only the rest of the body but also the wider environment around it. Instead of regarding the mind as some kind of computer that processes information, they see it as arising, at least in part, from the body's self-regulating operations – from the body's attempts to maintain a dynamic equilibrium in the face of changing environmental conditions. In other words, they argue that the higher order cognitive operations of the mind emerge out of the conflicting desires that attract the body towards some objects and repel it from others (Noë 2009, 67–95; Johnson 2018, 36–38). Such a vision of the mind implies precisely what Boyle so keenly understood – that the body's likes and dislikes are constituents of thought, effectively giving rise to the supposedly rational operations through which we come to grips with the things around us. Rather than being distractions from thought, these feelings are the substance out of which thought is made.

Bibliography

Adams, M. (2014) 'The Wax and the Mechanical Mind: Reexamining Hobbes's Objections to Descartes's *Meditations*', *British Journal for the History of Philosophy*, 22(3), 403–424.

Anstey, P. (2000) *The Philosophy of Robert Boyle*. London: Routledge.

Anstey, P. (2011) *John Locke and Natural Philosophy*. Oxford: Oxford University Press.

Aquinas, T. (1920–1922) *Summa Theologica*. Translated by Fathers of the English Dominican Province. 2nd ed. London: Burns, Oats and Washbourne.

Aristotle (1996) *Physics*. Translated by R. Waterfield and introduced by D. Bostock. Oxford: Oxford University Press.

Boyle, R. (1666) The Origine of Formes and Qualities. Oxford. Printed by H. HALL Printer to the University, for RIC: DAVIS.

Brooke, J. H. (1991) *Science and Religion: Some Historical Perspectives.* Cambridge: Cambridge University Press.

Clericuzio, A. (1990) 'A Redefinition of Boyle's Chemistry and Corpuscular Philosophy', *Annals of Science*, 47, 561–589.

Daston, L. and Galison, P. (2007) *Objectivity.* New York, NY: Zone Books.

Daston, L. and Park, K. (2001) *Wonders and the Order of Nature, 1150-1750.* New York, NY: Zone Books.

Dear, P. (1995) *Discipline and Experience: The Mathematical Way in the Scientific Revolution.* Chicago, IL: University of Chicago Press.

Dear, P. (2008) *The Intelligibility of Nature: How Science Makes Sense of the World.* Chicago, IL: University of Chicago Press.

Dear, P. (2009) *Revolutionizing the Sciences: European Knowledge and Its Ambitions, 1500-1700.* 2nd ed. Princeton, NJ: Princeton University Press.

Descartes, R. (1998), *The World and Other Writings.* Edited by S. Gaukroger. Cambridge Texts in the History of Philosophy. Cambridge: Cambridge University Press.

Descartes, R. and Schoock, M. (1998), *La Querelle d'Utrecht.* Translated by T. Verbeek. Paris: Les Impressions Nouvelles.

Galison, P. (1984) 'Descartes's Comparisons: From the Invisible to the Visible', *Isis*, 75(2), 311–326.

Garau, R. (2017) 'Taming Epicurus: Gassendi, Charleton, and the Translation of Epicurus's Natural Philosophy in the Seventeenth Century', in S. Fransen, N. Hodson and A. Enenkel (eds.) *Translating Early Modern Science.* Intersections: Interdisciplinary Studies in Early Modern Culture 51. Leiden: Brill, pp. 233–257.

Garber, D. (2013) 'Remarks on the Pre-History of the Mechanical Philosophy', in S. Roux and D. Garber (eds.) *The Mechanization of Natural Philosophy.* Boston Studies in the Philosophy of Science 300. Dordrecht: Springer, pp. 3–26.

Gaukroger, S. (1992) 'Descartes's Early Doctrine of Clear and Distinct Ideas', *Journal of the History of Ideas*, 53(4), 585–602.

Giglioni, G. (2006) 'Mastering the Appetites of Matter: Francis Bacon's *Sylva Sylvarum*', in C. Wolfe and O. Gal (eds.) *The Body as Object and Instrument of Knowledge: Embodied Empiricism in Early Modern Science.* Dordrecht: Springer, pp. 149–167.

Gillespie, S. (2007) 'Lucretius in the English Renaissance', in S. Gillespie and P. Hardie (eds.) *The Cambridge Companion to Lucretius.* Cambridge: Cambridge University Press, pp. 242–253.

Hock, J. (2021) *The Erotics of Materialism: Lucretius and Early Modern Poetics.* Philadelphia, PA: University of Pennsylvania Press.

Hooke, R. (1665) Micrographia: or Some Physiological Descriptions of Minute Bodies. London. J. Martyn and J. Allestry.

Hunter, M. C. (2010) 'Experiment, Theory, Representation: Robert Hooke's Material Models', in R. Frigg and M. C. Hunter (eds.) *Beyond Mimesis and Convention.* Boston Studies in the Philosophy and History of Science 262, pp. 193–219; Dordrecht: Springer.

Ivanova, M. (2021) 'The Aesthetics of Scientific Experiments', *Philosophy Compass*, 16(3), 1–9.

Johnson, M. (2018) *The Aesthetics of Meaning and Thought: The Bodily Roots of Philosophy, Science, Morality, and Art*. Chicago, IL: University of Chicago Press.

Jones, M. (2006) *The Good Life in the Scientific Revolution: Descartes, Pascal, Leibniz, and the Cultivation of Virtue*. Chicago, IL: University of Chicago Press.

Mulligan, L. (1994) 'Robert Boyle, 'Right Reason,' and the Meaning of Metaphor', *Journal of the History of Ideas*, 55(2), 235–257.

Newman, W. (2006) *Atoms and Alchemy: Chemistry and the Experimental Origins of the Scientific Revolution*. Chicago, IL: University of Chicago Press.

Noë, A. (2009) *Out of Our Heads: Why You Are Not Your Brain, and Other Lessons from the Biology of Consciousness*. New York, NY: Hill and Wang.

Principe, L. (1995) 'Virtuous Romance and Romantic Virtuoso: The Shaping of Robert Boyle's Literary Style', *Journal of the History of Ideas*, 56(3), 377–397.

Principe, L. (2000) *The Aspiring Adept: Robert Boyle and His Alchemical Adept*. Princeton, NJ: Princeton University Press.

Roberts, L. (1995) 'The Death of the Sensuous Chemist: The 'New' Chemistry and the Transformation of Sensuous Technology', *Studies in the History and Philosophy of Science*, 26(4), 503–529.

Sargent, R. (1995) *The Diffident Naturalist: Robert Boyle and the Philosophy of Experiment*. Chicago, IL: University of Chicago Press.

Shapin, S. (1984) 'Pump and Circumstance: Robert Boyle's Literary Technology', *Social Studies of Science*, 14(4), 481–520.

Shapin, S. and Schaffer, S. (1985) *Leviathan and the Air-Pump: Hobbes, Boyle, and the Experimental Life*. Princeton, NJ: Princeton University Press.

Sloan, P. (1972) 'John Locke, John Ray, and the Problem of the Natural System', *Journal of the History of Biology*, 5(1), 1–53.

Swann, E. (2020) *Taste and Knowledge in Early Modern England*. Cambridge: Cambridge University Press.

Von Hoffman, V. (2016) *From Gluttony to Enlightenment: The World of Taste in Early Modern Europe*. Urbana, IL: University of Illinois Press.

Walmsley, P. (2003) *Locke's Essay and the Rhetoric of Science*. Lewisburg, PA: Bucknell University Press.

Wilson, C. (2007) 'Lucretius and the History of Science', in S. Gillespie and P. Hardie (eds.) *The Cambridge Companion to Lucretius*. Cambridge: Cambridge University Press, pp. 131–148.

Wilson, C. (2008) *Epicureanism at the Origins of Modernity*. Oxford: Oxford University Press.

Wragge-Morley, A. (2020) *Aesthetic Science: Representing Nature in the Royal Society of London, 1650-1720*. Chicago, IL: University of Chicago Press.

Yolton, W. (1970) *Locke and the Compass of Human Understanding: A Selective Commentary on the "Essay"*. Cambridge: Cambridge University Press.

4 Science as Revelation

The Alchymist … by Joseph Wright of Derby

Adrian Holme

4.1 Introduction

The Alchymist kneels in his laboratory at night, illuminated dramatically by the white light emanating from a large receiving flask that he faces. His raised eyes are focused upon a bluish jet of vapour venting from the flask. His expression appears to be one of rapture. His right hand extends towards us, palm down – a gesture perhaps to wait, to watch. His posture, his outstretched arm, his wild grey locks, long beard and robes remind us not of a scientist but of a religious figure at prayer – an Old Testament prophet, a Magus, or a saint.

All around we see his laboratory, somewhat cluttered, yet purposeful, not especially chaotic, with pieces of equipment, many vessels, large and small, including the largest silhouetted equipment to the right, plus various papers and a terrestrial globe. There are shelves behind with glass bottles, and here the other two people in the picture are to be found, two youths, with curious expressions – are they surprised or somewhat sceptical? The worldly expressions of the boys, their faces illuminated by a candle, contrast with the ecstatic expression of the Alchymist himself. This dramatic scene is a Gothic one, appropriately dark, with mediaeval architecture more befitting a church or temple than a laboratory, and behind the Gothic arch window with leaded glass, we catch sight of the full moon, partially obscured by cloud.

Joseph Wright of Derby painted this work (Figure 4.1) in 1771 and reworked it in 1795 (Bamford and Wallis 2017), giving it the long prosaic title: "*The Alchymist in Search of the Philosopher's Stone, discovers Phosphorous, and prays for the Successful Conclusion of his Operation, as was the Custom among the Ancient Chymical Astrologers*". This cumbersome title purports to tell us everything we need to interpret this work. But, as so often with Joseph Wright, behind the matter-of-fact description lie a host of interesting questions. The principal question is why, some 150 years after Francis Bacon, in 1605, had debunked and discredited the principles of alchemy (Bacon 2004), and when modern chemistry was established as a respected discipline, should Wright

DOI: 10.4324/9781003263920-5

Figure 4.1 Wright of Derby (1771/1775). *The Alchymist in Search of the Philosopher's Stone, discovers Phosphorous, and prays for the Successful Conclusion of his Operation, as was the Custom among the Ancient Chymical Astrologers*. Painting, oil on canvas, 127 × 101.6 cm. © Derby Museums Collection.

have chosen to paint this subject, and, especially, to put it in a religious context? Was it simply a record of an historical event, the successful preparation of phosphorous? Or was Wright telling us something more? Perhaps Wright's contemporaries were puzzled, as the painting failed to sell and was still in his possession in 1795.

The purpose of this chapter is to look more deeply into this particular work, firstly as an example of the addressing of the aesthetics of scientific experiment by an artist; secondly, to explore the iconography, religious and alchemical, employed by Wright in this and other works, and the significance of the experimental apparatus; and, thirdly, to inquire into the interpretations of the painting and the meaning it may still have for us today, two and a half centuries after it was painted. Does it still carry a message about the nature of our own encounters with science and the fruits of science in technology?

In the next section, we will introduce Joseph Wright and consider his aesthetic, including some of the philosophical influences that may be found in his works, focusing especially upon his approach to light and the Candlelight paintings.

4.2 The Painter of Light

Joseph Wright was born in Derby in 1734. Although he spent time in London, Liverpool, and Italy, most of his life was spent in Derby, a town in the English Midlands, where he died in 1797. He was known as Joseph Wright of Derby in order to distinguish himself from another painter of the same name (Egerton 1990).

Wright is best known for his "Candlelight" paintings, influenced by the dramatic style of Caravaggio (1571–1610).[1] He has justifiably been called a 'painter of light' (see Nicholson's 1968 two-volume work) and also, in a complementary sense, a 'painter of darkness' (Craske 2022). The Candlelight paintings deal with nocturnal interior scenes in the houses of the rising middle classes and include some of Wright's most famous "scientific" works such as *A Philosopher Giving that Lecture on the Orrery, in which a lamp is put in place of the Sun* (1766), and *An Experiment on a Bird in the Air Pump* (1768). Wright also painted nocturnal scenes of the burgeoning industry of Britain's industrial revolution (he was associated with, though not a member of the Birmingham-based 'Lunar Society' in which scientists and industrialists met and discussed at meetings held near the full moon (Nicolson 1968) to facilitate travel). Wright's output included a series of paintings of blacksmith's forges at night (see below), as well as nocturnal landscapes. The Candlelight works cemented Wright's reputation at home and internationally – one of the forge paintings was purchased by Catherine the Great of Russia. Wright also painted portraits, plus other landscapes and allegorical works.

Wright's painting career took place mostly during the European Enlightenment, with its emphasis on the role of the rational, but there are many "Gothic" elements in his works – the nocturnal scenes, the Gothic architecture, the full moon – and so the works also foreshadow the Romantic movement which gathered pace in the latter part of his life. In fact, the interplay between rationalist "Enlightenment" aspects and more Romantic aspects such as heightened emotion, drama, and death are not untypical of Enlightenment artists (notably Goya for instance). In paintings such as *The Experiment on a Bird in the Air Pump* Wright explores, and appears to revel in, a creative tension between reason and the imagination as he also explores the tension between light and darkness.

Wright's treatment of light in the Candlelight paintings differs from the dramatic chiaroscuro technique associated with Caravaggio in that he characteristically incorporates a fairly central source of light, often occluded, in these paintings (in Caravaggio's works the light source lies outside the frame of the picture). In fact, as we shall see, there are often two sources of light in Wright's nocturnal paintings, one of which is the full moon, and this complicates an oversimplistic reading of the works in terms of light and dark.

4.2.1 John Locke

But before we leave the consideration of the overall relationship of light and dark in Wright's works, we should consider, as several writers have (e.g., Daniels 1999; Craske 2022), the influence of the writings of John Locke. In *An Essay Concerning Human Understanding* of 1689 (Locke 2004, the philosopher compares understanding to the working of the eye, which lets in light while taking "no notice of itself". For Locke, humans are initially a kind of passive blank slate, a *tabula rasa*. Sensations are passively received, like light entering the eye, upon which the mind reflects and learns. Locke was one of the first philosophers to reflect in a modern psychological way upon the process of visual perception, with his speculation on the way in which a blind person given sight might perceive forms visually. He concluded that we require experience in order to learn to interpret visual forms.

It is useful to understand Wright's Candlelight works, and indeed the paintings of the erupting volcano Vesuvius, in this way, as a process of education achieved by light entering a dark chamber and somehow hollowing out a space. As Paulson (1975) comments:

> Light has, from the start, shaped space for Wright. He seldom indicates the architectural structure of rooms; he defines space as a darkness hollowed out by light, and so basically a cave-like area.
>
> (Paulson 1975, 195)

The *Three Persons viewing the Gladiator by Candlelight* (c1764–1765) can be interpreted as a complex exploration of representation and visual perception in this Lockean sense. Three men in a darkroom, seated around the brightly-illuminated classical sculpture of the Gladiator, engage in looking at the form, as well as at a drawing of the statue (a two-dimensional representation of a three-dimensional representation, all represented within the two-dimensional painting). Later paintings of views from the inside of a grotto (e.g., the two versions of *Grotto with the figure of Julia* (1780)) may also be understood as examples of Lockean perception from within a dark chamber.

The influence of one more theory of John Locke – his concept of labour as the origin of economic value – will be explored later in this chapter when we consider Wright's paintings of the blacksmith's forges.

4.2.2 The Sublime

A further aspect of philosophy and art theory must be mentioned at this point in our attempt to understand Wright's treatment of light and dark. This is the theory of the "sublime". Originally meaning "raised" or "elevated", in the philosophy of Edmund Burke (1729–1797), set out in *A Philosophical Enquiry into the Origin of our Ideas of the Sublime and Beautiful* of 1757 (Burke 2004), sublime art, distinguished from "beauty", is associated with darkness, obscurity, confusion, and also terror:

> The images raised by poetry are always of this obscure kind .... But painting, when we have allowed for the pleasure of imitation, can only affect simply by the images it presents; and even in painting a judicious obscurity in some things contributes to the effect of the picture; because the images in painting are exactly similar to those in nature; and in nature dark, confused, uncertain images have a greater power on the fancy to form the grander passions than those have which are more clear and determinate ...
>
> (Burke 2004, 106)

The "sublime" was later to form an important part of Immanuel Kant's (1724–1804) aesthetic philosophy in which, again in contrast to beauty, it was associated with feelings of awe, especially in our experience of the power of nature.

> ... the irresistibility of [nature's] power certainly makes us, considered as natural beings, recognize our physical powerlessness, but at the same time it reveals a capacity for judging ourselves as independent of nature and a superiority over nature...whereby the humanity in our person remains undemeaned even though the human being must submit to that dominion.
>
> (Kant, quoted in Ginsborg 2022)

Kant's examples of the sublime include 'overhanging cliffs, thunder clouds, volcanoes and hurricanes' (Ginsborg 2022).

Thus, in Wright's Candlelight and other nocturnal landscapes, the darkness itself has not only symbolic functions, such as a Lockean metaphor for framing our understanding but also an aesthetic function, contrasting with beauty, of generating that feeling of pleasure in the viewer associated with the sublime.

4.3 The "Scientific" Works and the Scenes of Industry

To better understand *The Alchymist ...* it is useful to place this work in the context of the Candlelight paintings and the industrial scenes of blacksmiths' forges. It is instructive that it was first exhibited in 1771 alongside two paintings of blacksmiths' forges: *A Blacksmith's Shop* (1771) (Yale Center for British Art, Paul Mellon Collection) and another smaller *Blacksmith's Shop*, now lost (Egerton 1990). *The Alchymist ...* can also be understood in relation to two of Wright's most famous works that deal with "scientific" subjects, in the form of demonstrations: the aforementioned *A Philosopher Giving that lecture on the Orrery in which a lamp is put in place of the Sun* (1766) and *An Experiment on a Bird in the Air Pump* (1768). And it is these "scientific" works that we will now consider.[2]

If we include *The Alchymist ...* in these scientific works, we can see that Wright has chosen to depict subjects from three of the major branches of science of his day: astronomy, pneumatics, and chemistry. Of these three disciplines, all came to have significant practical benefits in the burgeoning Industrial Revolution of Wright's day: astronomy in navigation, pneumatics in the investigation of gases and the eventual development of the steam engine and the steam-powered forge, and alchemy in giving birth to a new chemical industry with applications in textiles and other industries (Bernal 1969).

4.3.1 The Orrery

A Philosopher Giving that lecture on the Orrery in which a lamp is put in place of the Sun (1766) (Figure 4.2) shows a working model of the solar system, which was, to 18th-century understanding, the universe. The source of light, a lamp put in place of the Sun, is hidden, as so often in Wright's paintings – a device that increases the dramatic and sublime impact of the painting. Its yellowish light illuminates not only the model of the solar system but also the eight people gathered around in the darkened room. Dominating the scene in the upper centre of the painting is the figure of the philosopher-scientist dressed in striking red robes, who is glancing towards the figure to his right, a younger man taking notes – Wright's friend and collaborator, the surveyor Peter Burdett. Spiralling around the table, we see a young woman looking pensively at the model, a boy almost entirely silhouetted whose body obscures the lamp, then,

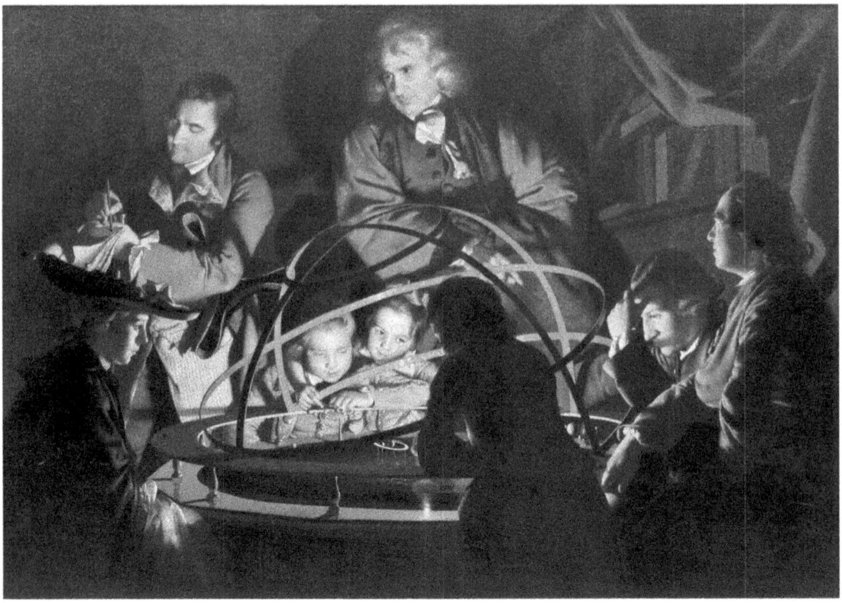

Figure 4.2 Wright of Derby (1766). *A Philosopher Giving that lecture on the Orrery in which a lamp is put in place of the Sun.* Painting, oil on canvas, 147.3 × 203.2 cm. © Derby Museums Collection.

on the far right, an older man looking towards the philosopher and a younger man looking downward in a thoughtful pose. Finally, at the centre, below the philosopher, we have two young children, a boy examining the moons of the planet Saturn, and a girl whose arm around the boy's shoulder points to one of the moons. To our right in the darkened room, we see some green curtains and dimly-lit bookshelves with some rather disorderly books.

On the one hand, this scene represents in a straightforward prosaic manner, as the title indicates, a lecture on the orrery, and that particular lecture, as Craske (2022) points out, involving replacing the central globe with a lamp, in order to practically demonstrate eclipses – which, we can assume, is what the philosopher is discussing. But, in terms of the deeper symbolic significance, the Orrery is a diminutive version of the mechanical universe described by Galileo (1564–1642), Descartes (1596–1650), and Isaac Newton (1643–1727) – a mechanical universe in which the relationship of "man" and God had shifted dramatically. Today, it is hard to fully appreciate the profundity of the philosophical shift that took place firstly with the Copernican "revolution" which placed the Sun at the centre of the cosmos rather than the Earth, overturning Catholic dogma (Galileo's endorsement of heliocentrism led to his being threatened by the Inquisition). And, secondly, because with his recently-invented telescope, Galileo discovered around 1609 that the moon was not unblemished but pock-marked and that Jupiter had moons of its own (Dampier 1948; Burtt 1967). These discoveries demolished the Aristotelian cosmology that had inspired both Christianity and Islam. The Heavens were no longer a separate realm but a continuation of the space in which the Earth resided, in which similar phenomena or Universal Laws operated. This new view, however, raised the troubling question "where was God?" and also, if the cosmos operated according to the rational cause and effect operation of mechanics, "what was the role of God after He had started up this great machine?"

But, the *Lecture* ... also features people, an apparently carefully selected range of people of different ages and both genders, somehow representing society (or at least the middle classes depicted here). To some degree, such selections of people might reflect the genre paintings of the "ages of man" (see for example paintings of *The Four Ages of Man* by Nicolas Lancret 1690–1743). But Wright's works are typically multi-layered, and in the *Lecture* ... there is a further dimension. As we progress around the table the heads of the people are illuminated by this artificial Sun in a way not dissimilar to the illumination of a planet or a moon – on the far right we even have a figure with a discernible crescent of light on his face. So, why might Wright be making a parallel between the rational planetary system and the group gathered around the Orrery? As Bernal (1969) argues, it was important for such connections to be found:

> Those [philosophers] of the eighteenth century ... could take the scientific world-picture which Newton had given them for granted. Their task was to extend it and to reconcile its findings, and still more its attitude of mind, with the new political and economic pattern that was beginning to appear in their own time.
>
> (Bernal 1969, 516)

Furthermore, Daniels states that comparisons could be drawn between Newton's gravitation and connections operating in the social world:

> The Newtonian theory of universal gravitation, as demonstrated on this cosmic scale, was used to articulate a range of social, spiritual, and political concerns. The faces of the figures around the orrery reflect the light at different angles, in a progression like the phases of the moons and planets they observe, establishing a connection between the social and cosmic worlds that Anglican theologians saw in Newton's principle of attraction.
>
> (Daniels 1999, 37)

Finally, as Paulson (1975) observes, the strength of illumination of the faces appears related to their ages – with the youngest most strongly illuminated (a device also apparent in *An Experiment on a Bird in the Air Pump*).

The subject of the orrery was clearly chosen carefully by Wright, not just for its philosophical and scientific significance, but also because in the Britain of the 18th century, the developments in astronomy underpinned improvements in navigation (Bernal 1969) – and this in turn facilitated trade of all kinds, including the hugely profitable Transatlantic slave trade and importation of commodities from the New World, as well as the policing of trade by the British Navy.

This painting thus functions simultaneously on various levels. Prosaically it is a representation of a plausible event, of a philosopher using the orrery to present a lecture on the workings of the solar system and especially eclipses (with individuals probably identifiable to an educated contemporary audience). It is imbued with the wider significance of the rational mathematical and mechanical cosmos, the social implications of 'gravity', even suggestions of Masonic influence (Daniels 1999), with Newton's God as the divine architect. It functions aesthetically, employing sublime effects of light and dark, and the symbolism of light in a dark chamber in a Lockean sense, commensurate with Locke's passion for education. It is a rich multi-layered approach used in the other scientific works as we shall see.

4.3.2 The Air Pump

The second of Wright's "scientific" works is almost certainly his best-known today. *An Experiment on a Bird in the Air Pump* (1767–1768) is an important work in London's National Gallery and has been widely reproduced (Figure 4.3). Today, the subject might seem merely a curious, bizarre, and cruel experiment, but it is hard to overstate the importance of the air pump to the history of science and the subsequent developments from the 17th century, culminating in theories of gases, advances in chemistry, understanding of physiology, and the development of the steam engine (Bernal 1969).

For more than 2000 years in Europe, the nature of space had been the subject of controversy. In Classical Greece, Parmenides (515–460 BCE) 'argued against the existence of empty space' (Dampier 1948, 21), though the common idea of philosophers was that space was full – a 'plenum' (Bernal 1969). Democritus (c460–c370 BCE) later argued that matter was composed of indivisible particles – atoms – that moved in the void or 'vacuum' of empty space, but in his *Physics* Aristotle (384–322 BCE) argued against the possibility of a vacuum, one reason being that it would allow a body to keep on moving, which would be absurd (Newton later crystalised this very absurdity in his first law of motion) (Bernal 1969).

When the debate about the vacuum resurfaced in the 17th century, it was also charged with challenging questions about the existence

Figure 4.3 Green, Valentine (1769) mezzotint engraving on paper (47.8 × 58.3 cm), after Joseph Wright of Derby, *An Experiment on a Bird in the Air Pump,* (c1767–1768) Painting, oil on canvas, 182.9 × 243.8 cm. © Derby Museums Collection.

and whereabouts of God. In a "real" and uniform mechanical universe as described by Newton and others, where was God to be found? Descartes' dualist philosophy divided existence into a material realm of physical extension, *res extensa*, and an immaterial world, the *res cogitans*, associated with the rational, the mind, and the "secondary", or sensual qualities of matter. Descartes denied the possibility of a vacuum in this realm of extension, as did Henry More (1614–1687) who insisted that a vacuum was not possible because divine extension would fill such a void (Burtt 1967).

When Otto von Guericke (1602–1686), a German scientist and mayor of Magdeburg, attempted to produce a vacuum by designing an air pump, he spent £4,000 – an enormous sum – on the project. He achieved vacua in large vessels and decisively demonstrated the power of his vacuum to the Holy Roman Emperor with his Magdeburg sphere – two hemispherical halves of an evacuated metal sphere, which required two teams of eight horses to pull apart (Bernal 1969).

Like the *Lecture ...* on the Orrery, *An Experiment on a Bird in the Air Pump* takes place within a domestic interior with a group of people seated and standing around a table. Light emanates from a candle that sits behind a liquid-filled glass vessel containing an oddly-shaped object and a glass rod. At the very centre is a scientific apparatus – the air pump, a table-mounted device with two cylinders (see descriptions in Nicolson 1968; Craske 2022). Above the pump stands a glass vessel, and within this, towards the bottom, is a white bird with an outstretched wing. Clearly, the air has been pumped out of this vessel and the poor bird can neither fly nor adequately breathe.

Ten people are present in this scene. Dominant is the central figure of the lecturer or demonstrator[3] who has one hand raised above the vessel, potentially to open the stopcock to allow air back in to save the bird's life. In a dramatic scene, heightened by the sublime effect of the lighting, this central character is particularly dramatic. An older man with side-lit face, long flowing grey locks, wearing a red robe, his gaze is directed not to others in the scene, but to us. The hand not raised is also open to us, somehow inviting us to participate in the drama.

Others in the scene range from two young lovers[4] on the left who appear more interested in one another than the spectacle of the experiment. Next to them, a boy gazes intently at the experiment. Then, largely in silhouette sits a man who holds in his hand a watch, evidently to time the experiment. On the right-hand side sits an old man looking downwards, his hands clasped around his stick, immersed in thought. To the right of the apparatus, we find a group of three: a young girl, her face bathed in light looks up at the bird, with a concerned yet fascinated expression; an older girl is hiding her eyes from the struggling bird; behind them an older man, presumably their father, has a comforting arm around the older girl's shoulder. He speaks to her and his right-hand points upwards to the glass vessel. Finally, over to the far right, away from the table, a boy, presumably the demonstrator's assistant (his physical distance presumably also reflects his lower social status), holds a cord, attached to which, high in the room, hangs the bird's cage. The assistant stands by a window through which we see, conspicuously, the full moon.

On the table itself are a number of objects commensurate with the delivery of a lecture on pneumatics and also the physics of breathing. These include a pair of Magdeburg hemispheres. The candle could also be part of the demonstrations, with a flame dramatically extinguished by the evacuation of air. The most curious object is that within the cloudy liquid in front of the candle. What is it? Many writers (e.g., Egerton 1990; Daniels 1999; Stafford 1999) have described it as a skull, or a diseased skull, a *vanitas*. However, on close examination of the painting in the National Gallery, London, it is very difficult to reconcile this amorphous mass, and some kind of rod, with a diseased skull. Far more convincing is Craske's suggestion (2022) that this is flesh of some kind – for a demonstration of the amount of air contained in flesh – or, perhaps most convincing, the suggestion by Donald (2007), taken up by Powers (2017), that this is a 'Lungs Glass', a bladder that could be used in place of a live animal. This fits well with the presence of other objects like the Magdeburg hemispheres that might be used by a demonstrator, and Powers cites in support of a sketch of the painting made by Wright (Derby Museum and Art Gallery) depicting what appears to be a "Lungs Glass" in a glass vessel.

There are many similarities to *A Philosopher Giving that lecture on the Orrery* …. A group of people, middle-class society in microcosm, of both genders and varying ages, give differing responses to a scientific spectacle. As Donald (2007) and Craske (2022) observe, there appear to be age-related and gendered differences in the responses of the audience. The most emotional response is seen from the girls in the group (though Wright is not bound by simple stereotypes and elsewhere depicts girls' cruelty to "nature" or to another girl – e.g. *Two Girls decorating a kitten by candlelight*, 1768–1770, or *Two Girls and a Negro Servant*, 1769–1770).

As in the *lecture* … we have a multi-layered response to the subject that permits several interpretations, without intending to fix one single interpretation (this is surely key to Wright's continued importance as an artist). And we have the prosaic descriptive and long title – *An Experiment on a Bird in the Air Pump.*

Like the orrery featured in the *lecture* …, this is old science. One hundred years earlier, in 1670, Robert Boyle (1627–1691) had vividly described his use of an air pump to evacuate air from a vessel in which had been placed a lark:

> … we pumped out of the receiver ¾ of the air that was in it before; then heedfully observing the bird, we perceived it pant very much, so that a learned physician (from whom I yet dissented), judged those beatings to be convulsive; having continued this for a little more than a minute and a half; the bird fell in to a true convulsive motion, that cast it upon the back. And although we made great haste to let in the air, yet before the expiration of the second minute, and consequently in less than half a minute from the time immediately preceding the convulsion, the lark was gone past all recovery, though divers means were used to effect it.
>
> (Boyle quoted in Todd 1946, 11)

Why is Wright, 100 years later, painting a scene of this experiment? Clearly, the scene was still potent and the subject of demonstrations to a lay public, as depicted here, as well as to specialists.

Similar to the *lecture* … we have the dramatic and sublime effects of the occluded source of light in the darkened room – with its Lockean implications. The brightest light falls upon the youngest. There is also, in a more obvious way than the *lecture* …, a use of religious iconography. The choice of a bird for this experiment is logical for historical reasons and also because a bird depends upon air for flight as well as breath. But it is no accident that Wright has chosen a white bird, an expensive cockatoo (Egerton 1990). This choice not only heightens the drama – the unlikely use of a valuable bird, perhaps a household pet, but in terms of religious iconology it resembles the *Paraclete*, the Holy Spirit, conventionally symbolised by a white dove, in a way surely obvious to an educated 18th-century audience. The high central positioning of the white bird only reinforces this connection (e.g., see the placement of the *Paraclete* in *The Baptism of Christ* by Piero Della Francesca in the UK National Gallery). Furthermore, the hand of the father figure explaining to the two girls points directly upward to the vessel and the bird with a gesture that recalls that of Saint John the Baptist in the painting by Leonardo da Vinci, who points heavenward, to Christ. Considering that the idea of the vacuum had been embroiled in religious as well as scientific controversy – was it possible to have a true vacuum that might imply an absence of the divine? – the painting could be suggesting that the air pump, and the new scientific picture that went along with it, might imperil not only a poor bird but perhaps the divine itself.

So, in this restaging, in educative and theatrical form, of an historical experiment by Robert Boyle, a drama is enacted that turns upon the withdrawal of air from a vessel containing a live bird. The demonstrator, who catches our gaze, seems to be inviting our involvement. Will the air be let into the vessel in time to restore the poor bird? In front of the apparatus and the source of light, there is an empty place at the table – for us, the viewers of the work. And, as we have seen, a layer of religious iconography adds another dimension. Will the Holy Spirit itself (in Greek the word πνεῦμα – pneuma – means both breath and spirit) be threatened by the vacuum?

4.3.3 Alchemical Readings of An Experiment ...

We could conclude our discussion of *An Experiment* ... at that point. But another meaning of the white bird seems to have generally escaped attention in the literature on Wright. For images of white doves in glass flasks are found in alchemical literature, where the white dove represents alchemical processes – 'The eternal process of transmigration or sublimation' (Klossowski de Rola 1973, 108), or, more precisely, 'the mercurial "spirit" that carries out the work' (Roob 1997, 131). And there is a further potential alchemical interpretation in Wright's painting. This is the presence of the full moon (partially obscured) in the window on the extreme right of the painting, to which our attention seems deliberately drawn in the composition by the presence of the boy. The full moon appears in Romantic painting and literature, where it represents nature, the female, and the "mercurial" processes of change. But alchemically, the moon also represents mercury – today understood as merely an element, but in alchemy, *Luna*, one of the polar forces of the cosmos (see Roob 1997).

In alchemical theory the other of the "dual principles" was sulphur, or *Sol*, which was male, dry and associated with the Sun. While the silvery mercury represented irrational nature, the yellow sulphur was associated not only with fire, Sun and Apollo but also with Reason. The interactions of these two principles, *Sol* and *Luna*, sulphur and mercury, were at the heart of the work of alchemy in perfecting the Philosopher's Stone – which was a process of spiritual purification as well as enabling the transmutation of metals – culminating in the incorruptible gold (Dampier 1948; Klossowski de Rola 1973; Roob 1997).[5]

We can therefore add this alchemical interpretation to our understanding of *An Experiment on a Bird in the Air Pump.* Sulphur, symbolised by the yellow light of the candle, is working with Mercury, symbolised by the moon, in an alchemical process (sublimation, symbolised by the white bird in the glass vessel). To express it another way: reason and nature are interacting in the course of the work.

Much like the *lecture ...*, *An Experiment on a Bird in the Air Pump*, has multiple interpretations: a dramatic play on the life and death of a bird in a vessel evacuated of air (despite Craske's (2022) scepticism it seems clear that it is on this that the drama turns – though the bird will almost certainly be released); the diverse social responses to a scientific demonstration, and a collision between the detached rationality of science and the emotions of people; a play on the religious significance of the vacuum with the bird representing the Holy Spirit, seemingly threatened by this scientific experiment; and, last but not least, the iconography of an alchemical process within a modern scientific setting.

There is one more aspect shared by the *lecture* ... and by *An Experiment* .... This is the centrality of the apparatus. In each painting, the apparatus, the orrery, the air pump takes centre stage. In each case, people are gathered around the apparatus, and in each case the apparatus appears as a kind of performer with a certain agency. We are moving, by the 18th century, from the age of 'tools' that surround the worker to an age of 'machines' surrounded by people (Von Amelunxen 2006). The machine and the technical apparatus are taking centre stage.

4.3.4 The Two Lights

Understanding these alchemical references makes sense of something of an enigma in the paintings of Wright. This is his use, time after time, of two sources of light within his paintings – one a yellow sulphurous light and the other the full moon – always partially obscured by something whether a cloud or foliage or a window lattice. Although this convention does appear occasionally in the work of other artists (de Loutherberg's *Coalbrookdale by Night*, 1801, is a notable example) it is to be found in many of Wright's works. Overall, the use of the two lights of *Sol* and *Luna* appears to be associated with some kind of "progressive" change, scientific or industrial (such as the Blacksmith's forges – see below), or in the case of the *Earthstopper on the Banks of the Derwent* (1773), or the several versions of *Cottage on Fire* (1787–1790), some supposed "improvement" in the land, through fox hunting or the clearances of peasantry. In the case of the many *Vesuvius* paintings in which the light of the full moon complements the sulphurous light of the volcano, there is a kind of natural alchemy at work – a change in the land wrought by nature herself.

This alchemical symbolism of the two lights becomes even more evident when we note that the alchemical literature sometimes features images in which both *Sol* and *Luna* occur in the same image (e.g., Figure 4.4).

We will now explore in more detail the use of the two lights and the alchemical as well as religious iconography in Wright's series of paintings of blacksmiths' forges. These paintings pose a question as to whether the burgeoning capitalism and industry of Wright's day might be considered as a new kind of alchemy.

Figure 4.4 Maier, Michael (1618) 'The Sun needs the Moon as the cock needs the hen'. *Atalanta fugiens*. Emblem XXX. In: Klossowski de Rola, S. (1988) *The golden game: Alchemical engravings of the seventeenth century*. London: Thames and Hudson.

4.4 The Blacksmiths' Forges

From the mid-1760s to the early 1770s, Wright painted not only the three "scientific" works but also, between 1771 and 1773, a series of five depictions of blacksmiths' and iron forges at night. In 1770–1771 he painted two versions of *A Blacksmith's Shop*: one now at Yale Center for British Art (Figure 4.5), the other at Derby Museums and Art Gallery.

Figure 4.5 Wright of Derby (1771) *The Blacksmith's Shop*. Painting, oil on canvas, 128.3 × 104.1 cm. Yale Center for British Art, Paul Mellon Collection. Copyright Status: Public Domain.

These similar compositions show three blacksmiths working to produce a horseshoe at night. In the Yale picture, a group of people to the right (recomposed for the Derby painting) show differing reactions to the hammering of the glowing iron, which throws its light over the smiths and illuminates the scene. Above, to the right, the full moon (partly obscured) is visible. To the left, a group of people and a horse being shod, or waiting to be shod, are just visible by candlelight.

Wright's notes from the late 1760s for such a painting are typically matter-of-fact (Paulson 1975; Egerton 1990):

> Two Men forming a Bar of Iron into a horseshoe – from whence the light must proceed ... Horse Shoes hanging upon ye walls, and other necessary things, faintly seen being remote from the light – Out of this Room, shall be seen another, in wch. a farier may be shoeing a horse by the light of a Candle. The horse must be Sadled and a Traveller standing by The Servant may appear with his horse in his hand – on wch. may be a portmanteau – This will be an indication of an Accident having happen'd, & shew some reason for shoeing the horse by Candle Light – The Moon may appear and illumine some part of the horse if necessary –
>
> (Joseph Wright quoted in Egerton 1990, 98)

Wright is clearly concerned with the light that must "proceed" from the bar of iron, but the reference to the moon appears almost incidental. Yet the full moon is visible in both versions of *A Blacksmith's Shop*. What is more, the composition in the Derby version of the painting has been adjusted compared with the Yale version in an interesting way. In the Yale painting (Figure 4.5) the noble-looking young blacksmith wields his hammer such that it aligns with and directs our sight to the iron bar. In the Derby painting, the hammer now aligns directly as to connect, even more clearly, the glowing iron, the smith, his hammer, and the full moon behind.

Other interesting features of these paintings include the surrounding building – apparently a ruined church or abbey with carved angels visible above the scene (Egerton 1990). This, and the other three paintings of forges bear resemblances of composition to nativity scenes with, as Paulson (1975, 191) writes, 'the half-open and ruinous building, the country folk, even the farm animals, and the unearthly glow at the centre' (in a traditional nativity, light would emanate from the infant Christ). In Wright's paintings of blacksmiths shops 'the Holy Child is replaced by a piece of molten metal' (*Idem*). Paulson also notes that in this and other paintings something is being done to nature. He also observes, correctly, that the light from the hot iron is unlike the light of a candle (though candlelight is also visible to the left in the Derby and Yale paintings) and contrasts with the natural light of the moon.

In in a similar fashion to the "scientific" subjects, clear religious references, here to the Christian nativity, are complemented by further references to alchemy. In this regard, Solkin (2003) relates a fascinating finding concerning the Yale *Blacksmith's Shop*:

> At some point in the summer or autumn of 1770 Joseph Wright of Derby worked a small horizontal piece of gold leaf onto the lead-white underlay of a picture he was painting of three blacksmiths crafting a horseshoe at night. He then covered up this narrow strip of metal with an opaque layer of Naples yellow, which he had fashioned into the likeness of a glowing iron ingot ...
>
> (Solkin 2003, 167)

Although there may be prosaic technical reasons why Wright might have wished to embed gold under his depiction of an iron ingot, this curious act obviously invites alchemical interpretations. Solkin points out that the Yale painting was first exhibited in 1771 alongside Wright's *Farrier's Shop*, and *The Alchymist* .... Wright, he suggests is stating that he himself is a metal worker – lead is contained in the lead white, while 'Naples yellow contains cadmium, iron and titanium' (Solkin 2003, 189). Wright, suggests Solkin, is aligning himself with 'an industrious craftsman, whose "ingenious labor" gave moral, economic and social value to the products of his brush' (*Idem*). He concludes that there is a kind of alchemical process at work within Wright's work as an artist:

> Within the core of the burning ingot, and at the symbolic heart of Wright's notion of artistic labor, lies a quasi-magical act of material transformation – the transmuting of gold into iron. This sleight of hand endows painting with the character of alchemy (albeit in reverse), as the artist himself must have known full well. ... he uses forging – as he does alchemy – as both a metaphor and a platform for his "very great and uncommon genius" [quoting Wright] for wresting luminosity from the raw materials of paint.
>
> (Solkin 2003, 191)

Solkin points to the influence upon Wright of John Locke's theory presented in his *Second Treatise of Government* of 1690, that it is labour that is 'the origin of all economic value' (*Idem*: 174). He shows that Wright's depictions of the smiths are not demeaning as in previous works by other artists. In Wright's work:

> Labor makes wood and stone into buildings, refines raw ore into iron, crafts wood and iron into hammers and bellow and drums and watermills, and changes iron itself into horseshoes and all manner of necessary goods; at the centre of the *Blacksmith's Shop* [1770-1]

> we see a hot, malleable ingot, a piece of raw matter in the process of being hammered into useful form. This labor gives meaning and purpose to nature, while supplying the foundation of property, as well as the mechanism for producing civilization – and art.
>
> (Solkin 2003, 174)

So, here we have labour as a transformative process akin to alchemy – the labour of the blacksmiths, transforming nature, and the labour of the artist himself effecting a kind of alchemical transformation in his materials. In these terms, perhaps this is not "alchemy in reverse" as Solkin suggests, but a new kind of modern alchemy made possible by this coming together of science, technology, and capitalism in the "Industrial Revolution"[6] with labour at its heart. Wright's Yale painting, with its hidden gold leaf, is not about transforming gold into iron, but exposing the latent potential for finding gold within the humble material of iron, under the lead paint. Similarly, the labour of workers and craftspeople could transform humble materials like iron into actual gold – putting real gold into the pockets perhaps of the worker but more so the rising capitalist class whose investments reaped rich rewards.

This metaphorical modern alchemy still involved the working together of these two polar forces of *Sol* (Sun, sulphur, male, Reason) and *Luna* (Moon, Mercury, female, nature) – in other words Reason and nature had to work together, with labour effecting this alchemical work. This would explain the clear symbolism of the blacksmith's hammer in the Derby painting – connecting the lunar principle with the material being transformed. Furthermore, there are in fact not two, but three sources of light in the Yale and Derby paintings – the mercurial, full moon, the sulphurous yellow candlelight (to our left) and the shining new light of the iron ingot; the alchemical transformation of nature, yielding gold through the miraculous working of technology. The light from the centrally positioned glowing ingot that has the attention of the smiths and the spectators around it, reminds us of spiritual revelation; a new light in place of the light traditionally emanating from the Holy Infant in a nativity, an eerie light from a modern industrial nativity. This is also an alchemical nativity, since it is the working together of *Sol* and *Luna* that can produce the Philosopher's Stone (*lapis philosophicus*) or 'Universal Quintessence' (Klossowski de Rola 1997) that, in Christian alchemical tradition, was equated with Christ (see Roob 1997) – the agent of both material and spiritual purification.

Commenting on Wright's use of iconography in the iron forge paintings and the *Orrery*, Ronald Paulson concludes that 'Wright makes overt the fact that his subject is the secularisation of a miracle' (Paulson 1975, 191). He suggests that Wright 'does not appear to be switching his allegiance to the scientific source of light, but rather opposing both the religious and the scientific to the natural so that all three become part of a new construction' (*Ibid*., 191). This makes sense, for Wright is not removing the spiritual, or indeed alchemical, implications of these scenes, and his intention seems neither religious nor irreligious (Wright appears to have been a 'staunch' Protestant (Craske 2022, 134)). As with the gold, lying latent within the iron (literally so in the Yale *A Blacksmith's Shop*), religious feeling still lies latent within an industrial scene or a scientific experiment. Similarly, alchemical principles, this play with occult transformative forces that can transmute a humble metal to gold, still appear latent within a modern industrial or scientific context. Neither should we presume that Wright was intentionally obscure in his use of iconography. This religious and alchemical iconography presumably resonated with the educated audience of his work (see Vertesi 2007), as religious iconography is still resonant today, and can, for example, be identified within many a 21st-century cinema film or animation.

We can now return to a consideration of *The Alchymist* ... that we introduced at the beginning of this chapter.

4.5 The Alchymist

4.5.1 The Alchemical Legacy

There is no space here to go into alchemy in detail; however, some brief context may be helpful. The word 'alchemy' derives from the Arabic *alkīmiyā*, itself derived from the Greek *χημία*, *χημεία* – 'the art of transmuting metals' (Alchemy 1984). We owe the word "chemistry" to its alchemical roots. The origins of alchemy in Europe are thought to lie in Ancient Egyptian metallurgy as well as the likely connected alchemical traditions of India and China, and the European tradition derives particularly from Islamic and Judaic alchemy that drew upon Neo-Platonic philosophy. Although often crudely characterised as a search for the Philosopher's Stone that would enable transmutation of base metals into gold, and a related quest for the Elixir of Life, alchemy was as much a spiritual quest for purification as a material quest for wealth. In the cosmic view of the alchemist correspondences, hidden or occult influences, linked different levels of the cosmos – the greater macrocosm, with the stars and wandering planets, the sublunary and material world, and the human microcosm. In time alchemy gave birth to modern 'rational' chemistry as well as chemical industries, pharmacology, and improvements to brewing and distilling (for further introduction, see Klossowski de Rola 1973; Linden 1974; Roob 1997; Linden 2003; Holme 2014).

Wright's choice of an alchemist as the subject of a painting is a less obvious choice than the *lecture on the Orrery* ... or *An Experiment on a Bird in the Air Pump*, covering as they did two well-theorised and established areas of natural philosophy each of which had tangible benefits to industry. Unlike astronomy and physics, chemistry had struggled to find the underlying rational principles and theory that would set it upon a reliable mathematical basis (see Bernal 1969), though '... after the fifteenth century ... New substances with remarkable properties, such as phosphorous, were accidentally produced' (*Ibid.* 473).

The accidental discovery of phosphorous was credited to the German alchemist Hennig Brandt (1630–1710) (Nicolson 1968), around 100 years before Wright's painting, and phosphorous was later prepared by Boyle (Dampier 1948). However, even in Wright's day phosphorous was still understood as a compound rather than a fundamental "element" of chemistry. There is no indication that Brandt is depicted in Wright's work.

If we compare Wright's painting of the Alchymist with the other two "scientific" works, we see some important differences (and remember that Wright is not known to have considered these three works as a set, and would probably not have used the term "scientific" to describe them). Both *An Experiment on a Bird in the Air Pump*, and *That Lecture on the Orrery* ... take the form of demonstrations – consciously-constructed didactic spectacles with at least some theatrical aspects. In *An Experiment on a Bird in the Air Pump* the theory behind the "experiment" is well established, and only the outcome, the life of death of the bird is uncertain (though doubtless, the intention would have been a miraculous resurrection). But the "Alchymist", as the title informs us, makes a discovery of something new. His two assistants similarly do not know what to make of this appearance.

Another significant difference from the other two paintings is that the period in which the painting is set is uncertain. The *lecture on the Orrery* and *An Experiment on a Bird* ... are clearly set in Wright's day. But the *Alchymist* is not contemporary with Wright – the "discovery" of phosphorous had taken place 100 years before – and the Gothic architecture of the laboratory, with its Mediaeval associations, as well as the historical clothes of the Alchymist and assistants, give the picture an unspecified location in time.

4.5.2 *Alchemists in Their Laboratories*

There is a long tradition of depiction of alchemists in European art, not always complimentary. The 16th-century woodcut by Hans Weiditz (Figure 4.6) reveals alchemists as targets for satire. The two alchemists in their chaotic laboratory are surrounded by tools. One works his bellows (alchemists were disparagingly known as 'puffers' (Klossowski de Rola 1973)), while the other scratches his head gormlessly. Similarly, a 16th-century satirical work entitled *The Alchemist* (1558[7]) by Pieter Brueghel the Elder (c. 1525–1569) shows a family enrolled by an alchemist (reading his book) in a futile quest for gold and being reduced to penury as a result. These reflect a popular image of alchemists as charlatans, chaotic, and engaged in a misguided or delusory quest. Satirical

Figure 4.6 Weiditz, Hans (c1500–1536), *Zwei Alchemisten in der Werkstatt*, Woodcut, (96 × 157 mm). In: *Neue Künstliche, Wohlgerissene, vnd in Holtz geschnitene Figuren*. Vincentii Steinmeyer. Frankfurt a.M. 1620. Image © Kupferstich-Kabinett, Staatliche Kunstsammlungen Dresden.

depictions of alchemists persisted into the 18th and 19th centuries in the work of political cartoonists, such as James Gillray and Sir John Tenniel.

But there were also serious depictions. Wright himself owned a painting of an alchemist by the 17th-century Dutch painter, Thomas Wyck (or Wijck) (c1616–1677). Although the whereabouts of that particular painting is unknown (Bamford and Wallis 2017), Wyck produced a series of paintings of alchemists in their laboratory. *Interior with an alchemist*,[8] for example, shows some striking similarities to Wright's work in tones and colours. It is a sober depiction of an alchemist at his desk, surrounded by a collection of ceramic and glass vessels and jars. The space is untidy but a site of activity. As in Wright's painting, the alchemist is not alone – there are two others present, a man working with a large vessel and pestle, and a woman, perhaps a servant. The disorderly papers show some resemblance to the books in Wright's painting. Despite the similarities in the style and serious treatment, the striking difference is the lack of dramatic action in Wyck's work, compared with Wright's *Alchymist*.

There is another depiction of an alchemist that must inevitably have influenced Wright's painting (Daniels 1999). This is the engraving by Heinrich Khunrath (c1560–1605) of an alchemist praying in his laboratory from his

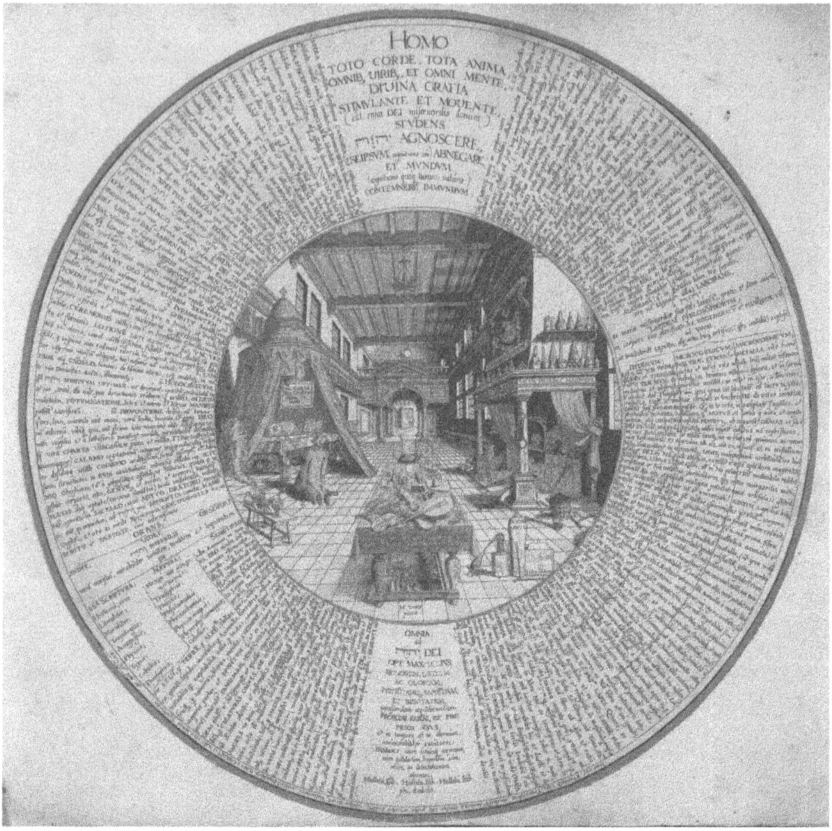

Figure 4.7 Khunrath, Heinrich. (1595). *Amphitheatrum sapientiæ æternæ, solius veræ*. Bookplate. By courtesy of the Department of Special Collections, Memorial Library, University of Wisconsin-Madison.

popular book *Amphitheatrum sapientae eaternae* (1595) (Figure 4.7). The alchemist's laboratory is here depicted as a site of work, with his alchemical apparatus and oven, and also as a site of prayer. The alchemist, seen from behind, prays, kneeling with outstretched arms, in front of his oratorium. The laboratory has a strong sense of order and discipline, in marked contrast with Wyck's more chaotic space. Musical instruments placed centrally on a table 'remind us that sacred music and harmony are supposed to accompany and define the work' (Roob 1997, 331).

Comparison with these two images is instructive. Wyck's painting is active, rather disordered, but not dramatic – suggesting ongoing serious, laborious work. Khunrath's engraving shows activity, spiritual and material, in a remarkably organised and harmonious space, and, like Wyck's image, is not dramatic. Khunrath's alchemist, over to one side, is a diminutive figure. There is no obvious "event", no historical "moment" in either depiction.

Wright, on the other hand, depicts a moment of history – reinforced by the clock on the wall specifically showing eight o'clock – a moment in time. Unlike the depictions by Wyck and Khunrath, Wright has endeavoured to make his work as dramatic as possible – implying a sublime suddenness of light and dramatic contrast with the darkness, with the ghostly blue jet erupting from the bright phosphorous in the receiver. We are witnessing a kind of miracle. The title informs us that this is a secular phenomenon, a discovery of science, yet the iconography is clearly religious, and the reaction of the Alchymist, dressed like a prophet or a saint, shows profound emotion, awe, reverence.

4.5.3 Scientific Credibility

Wright went to some trouble to give scientific credibility to his painting of the discovery of phosphorous. He sought advice from his friend Peter Burdett as well as from Dr Matthew Turner who lectured in chemistry, and Wright appears to have followed the account of the preparation of phosphorous by Pierre Joseph Macquer in his *Elements of the Theory and Practice of Chymistry* (Nicolson 1968; Bamford and Wallis 2017; Powers 2017). This account of the preparation of phosphorous from urine records that after many hours of boiling and various further stages including heating in a furnace, the phosphorous will come down from the furnace into the glass receiver, partially filled with water, at which time the phosphorous ignites spontaneously with a burst of white light as it combines with the oxygen in the flask. In Wright's painting, the prominently-placed glass balloon receiver supported upon its three spindly legs has a small hole through which comes the jet of phosphorous vapour that spontaneously combusts in the air emitting a bluish light.[9]

Burdett sent Wright his own depiction of the apparatus (held in the Derby Museum and Art Gallery) with some thoughts on the painting (see Figure 4.8). We can see that Wright followed the suggestion of the design and position for the furnace and the glass receiver, though the receiver lacked the hole in the balloon through which the bluish jet of phosphorous could escape. Wright also appears to have followed Burdett's suggestion of the setting within gothic architecture.

With Wright's attention to scientific credibility and his choice of contemporary scientific equipment (see Vertesi 2007), his painting presents an essentially modern chemical laboratory. Yet, many other details – the Gothic setting, the attitude of prayer, the apparently Jewish prayer shawl, the terrestrial globe above the Alchymist alluding to the astrological associations of the elements to the planets, the papers (on some of which are inscribed alchemical symbols (see Powers 2020)) – present an apparent contradiction to modern rational chemistry. Why did Wright seek to mix these modern and alchemical elements in such a conspicuous way?

Figure 4.8 Burdett, Peter Perez (1771). Sketch for *The Alchymist* in a letter to Joseph Wright of 4 February. Pen, ink, and wash on paper. Paper size 33.7 × 27.4 cm. © Derby Museums Collection.

4.6 Interpreting *The Alchymist* …

Wright presents his "scientific" subjects – the *lecture on the Orrery …*, *An Experiment on a Bird in the Air Pump*, and *The Alchymist* with an attention to technical detail and accuracy. The first two paintings are dramatic presentations that exemplify key scientific principles in a highly dramatic way, yet they remain plausible contemporary didactic demonstrations. We have seen that Wright was happy to utilise both religious and alchemical iconography to complicate the straightforward reading of the works – and we may assume that these iconographies could have been understood by his contemporary audience (Paulson 1975; Vertesi 2007). Of these three works, *The Alchymist* ... is the most puzzling and contradictory. We are not witnessing a dramatic, yet rational, controlled "demonstration" by an authority – a demonstrator or lecturer – a piece of scientific theatre. We are witnessing a genuine "discovery" – an eruption of something new and unprecedented, an historic moment, sublime in its suddenness and drama. Whether the Alchymist is as surprised as his assistants, we are not quite sure, because he has his own alchemical understanding of this event. Yet it appears as a miracle and is received as such, with prayer and awe. But the title also informs us that he is mistaken. This is not a miracle *en route* to the Philosopher's Stone, but the discovery of phosphorous. His prayers are evidently in vain. However, if they are, as the title implies, in vain, why is the Alchymist depicted with such apparent sympathy, with dignity, with authority?

There are two aspects to the interpretation of *The Alchymist* ...: the first is what Wright might have intended, and what it meant to the audience of his day; the second aspect is what the work might mean to us today – what we may still glean from Wright's aesthetic presentation of this scientific/alchemical experiment.

It is interesting to review the various interpretations offered in the literature. The most straightforward interpretation is that the painting means precisely what that long, prosaic title says: *The Alchymist, In Search of the Philosopher's Stone, Discovers Phosphorous and prays for the successful Conclusion of his operation, as was the custom of the Ancient Chymical Astrologers*. Craske (2022) takes this straightforward stance, drawing evidence from Wright's painting of *Miravan*, that *The Alchymist* ... is indeed about folly:

> ... an old man when faced with clear evidence of a natural wonder cannot appreciate it. Instead, this man prefers to believe that there may be a treasure greater than even nature can provide.
>
> (Craske 2022, 195)

This straightforward account tallies with Paulson's analysis of Wright's method as 'the secularisation of a miracle' (1975, 191) but does not explain the sympathetic treatment of the alchemist, who, far from looking foolish, resembles, in his fine robes, impressive beard and dignified bearing, a prophet, a Magus, or perhaps a saint. Nor does this take full account of the complexity of overlapping iconographies that we have seen in the "scientific" works and the iron forge paintings.

Nicolson (1968), noting Wright's accurate attention to the science of the method described by Macquer, saw *The Alchymist* as a modern scientist who had, nevertheless, produced something of value – an idea that can be traced back at least as far as Francis Bacon in *Novum Organum*, 1605 (2004), who wrote that even though the methods of alchemy were discredited, it had nevertheless yielded valuable results.

Daniels (1999) took a different view. The presumed influences of Wyck and Khunrath were both sympathetic to alchemy. And there were reasons to be sympathetic. Politically radical scientists, in those revolutionary times, such as Wright's physician Erasmus Darwin and the chemist Joseph Priestley 'cleaved to the notion of alchemy as a progressive, materialist religion, in opposition to the priestcraft of their day and it's cloaking of human greed and oppression'. Furthermore, argued Daniels, *The Alchymist* aligns with Wright's ambivalent or sometimes hostile attitude to the Royal Academy and 'it's ethos of separating the fine from the mechanical arts':

> Like the forge scenes Wright exhibited at the rival Society of Artists, *The Alchymist* dignifies a low subject, turns a technical operation into a noble pursuit. In Masonic terms it presents a 'chemical theatre' of material and spiritual transformation.
>
> (Daniels 1999, 30)

The suggestion of a sympathetic attitude to alchemy, aligned with radical political allegiances, helps explain some of the striking contradictions in Wright's painting: his sympathetic highlighting of a seemingly anachronistic and discredited practice, and his positive depiction of the central figure. And further support for this view comes from Vertesi (2007) who analysed the trinity of lights in the painting – the moon, the candle, and the receiving flask – in terms of Masonic concepts of the 'Three Great Lights', which were, in order of increasing potency, 'nature, reason and the divine (either in revelation or salvation)'. Wright would, she states, have been familiar with this symbolism, as would his patrons. The three sources of light visible are thus the faint moonlight associated with nature, the brighter candle associated with reason, and the brilliant light from the phosphorous with its pale blue jet, which is both a light of artifice and also symbolises the divine.

We can also describe this in similar alchemical terms – the mercurial moonlight and the yellow sulphurous flame of reason are both overshadowed by this light of revelation associated with preparation of the Philosopher's Stone. This light has a sudden and sublime appearance. The Alchymist experiences it as a kind of divine revelation, spiritually and alchemically, causing him to fall into prayer in front of a piece of scientific apparatus. There is a rational explanation, of course: the painting's title tells us explicitly in words. But, despite the title, the painting itself does not function primarily in that rational way. We witness a dramatic event, and the experience of the viewer is not unlike that of the Alchymist himself who is struck by this startling spectacle.

The symbolism of the phosphorous is of course ambivalent. It could be argued to be part of nature. Yet it is the appearance of something new – and therefore effectively artificial:

> ... phosphorous's analogical relationship to God makes the miracle of revelation part of the chemical scene. Just as divine revelation is a miracle ... the appearance of phosphorous is similarly miraculous. While man may indeed have created the artificial substance, the implication is that this discovery is no more humanly controllable than the visitation of theophany.
>
> (Vertesi 2007)

Vertesi concludes that:

> *The Alchymist* ... presents us with a hermitic discoverer of an artificial light, devoid of the content of reason required by philosophical practice, dominating other God-given or traditional sources of knowledge and indicating miraculous intervention into the laboratory.
>
> (*Idem*)

We can (though Vertesi does not do this) extend this analysis to Wright's paintings of blacksmiths' forges – in which the glowing ingot is also a third light, the light of artifice rather than nature, similarly equatable with divine revelation.

4.7 Conclusion

We have seen that, along with Lockean devices and sublime effects, Joseph Wright of Derby utilised religious and alchemical iconographies (which were to some degree linked) in his paintings of scientific and industrial subjects. We have also noted the centrality of the apparatus or equipment in the scientific works, which forms a focal point for the human subjects. As Vertesi (2007) points out, religious and alchemical iconographies would have been legible to Wright's public, even though religious and alchemical themes were not generally declared in the scientific and industrial works, and did not feature in the titles. *The Alchymist* ... is the notable exception in that the themes of religion and alchemy are explicitly present.

4.7.1 An Equivocal Modernity – Science as Revelation

Hannah Arendt (1998) observed that, in modernity, a distancing from religious content can be related to the survival and success of art:

> Even if the historical origin of art were of an exclusively religious or mythical character, the fact is that art has survived gloriously in its severance from religion, magic, and myth.
>
> (Arendt 1998, 167)

It is helpful to consider Joseph Wright in the context of Enlightenment modernity. Although there are historical and allegorical works, he was, unusually, a painter of modern subjects – generally shunning the Neoclassicism favoured by the art establishment of the Royal Academy (Vertesi 2007). Wright depicts the secular science and industry of a rational, progressive, Enlightenment age. But it is an equivocal modernity, for his paintings suggest that these religious and alchemical forces are still at work, though latent, under his framing of progressive modernity. The alchemical process in particular appears linked to some kind of progressive change, whether in science, in industry, or in the land – and is also akin to the progressive forces of capitalism, a new kind of alchemy creating gold in the form of wealth.

The Alchymist ..., as we have seen, is exceptional in that it explicitly addresses alchemical and religious themes. Yet, it still does so under a framing of modernity. In fact, if we accept Arendt's argument (2016) that modernity in the 18th century was associated with 'the breakdown of the old Roman trinity of religion, tradition, and authority', we might see *The Alchymist* as a quintessentially modern work. The painting's descriptive title tells us that the Alchymist is praying, yet clearly, his prayers are in vain; the tradition of alchemy has produced phosphorous but will not produce the Philosopher's Stone for which he prays; the authority of these beliefs no longer holds.

However, as with the other paintings, the message of progressive modernity is equivocal. An analysis of the iconography of *The Alchymist*, and especially the trinity of the 'Three Great Lights' (Vertesi 2007) – of nature, reason, and the divine, tells us that artifice, the sudden brilliant manifestation of light, and the eerie blue jet of vapour resulting from the application of science and technology has the appearance and impact of a theophany, a revelation of the divine. The *Alchymist* experiences it as revelation, but the title informs us that it is science. As so often in Wright's works, we, as viewers are given the choice – we may take it as a history painting of a discovery, or we may also experience something of the Alchymist's awe and reverence – the experience of scientific discovery as revelation.

Perhaps this understanding that artifice arising from science and technology may be experienced as divine revelation is the most significant message to the contemporary world of the 21st century. This technological manifestation of brilliant white light must have seemed something akin to a miracle to an audience in the late 18th century. But, today, our familiarity with bright light at the "at the flick of a switch" has perhaps dulled us to this feeling. Nevertheless, we are frequently confronted, directly, or through mediated spectacle, with phenomena that appear miraculous – at least at first encounter. As D. H. Lawrence wrote:

> Today, man gets his sense of the miraculous from science and machinery, radio, airplanes, vast ships zeppelins, poison gas, artificial silk: these things nourish man's sense of the miraculous as magic did in the past.
>
> (Lawrence quoted in Gray 2002, 123)

The problem with a miraculous experience of the 'artificial' fruits of science and technology, as Vertesi's (2007) analysis of the 'Three Great Lights' shows, is that those lesser lights of nature and reason (or *Luna* and *Sol*) are so much quieter, outshone by the brilliant appearance of revelation.

Today, in the 21st century we are confronted with problems that have resulted from the developments of science and technology, that are sublime in their enormity (such as climate change, environmental destruction, nuclear annihilation), or sublime in their dramatic visual spectacle (such as modern warfare). Perhaps a clearer understanding of our potential to experience artifice stemming from science and technology as akin to revelation, outshining those lesser lights of nature and reason, may help us pay more attention to those lesser lights – which are also *our* nature and *our* reason. For we need to be able to respond with reason, and with nature, to the seemingly miraculous science that continues to transform our world for good and for ill.

Acknowledgements

I would like to thank Derby Museums, and particularly Matt Edwards, Curator of Visual Art and the Joseph Wright Study Room, for their kind support in research for this chapter. I would also like to thank Jonathan Powers, Professor Emeritus of Derby University for sharing his insights into Wright's paintings. I thank the Library staff of the University of the Arts London for assisting with access to some of the publications on Joseph Wright of Derby. Finally, I am very grateful for the invitation and encouragement of the book's editors, Dr Milena Ivanova and Dr Alice Murphy with whom it has been a pleasure to work.

Notes

1 According to C. R. Hill (1975), Wright derived his influence from Caravaggio via the Utrecht School, and a Flemish influence is apparent also in Wright's choice of subject matter.
2 In Wright's day, the term 'scientist' was not in use as it is today, to describe the specialist working in science. Indeed, before Newton (1642–1726/27) 'science was simply natural philosophy, and the influential figures of the period were both the greatest philosophers and the greatest scientists' (Burtt 1967, 17). Hence, the 'Philosopher' giving the lecture on the Orrery might today be called a 'scientist'.
3 For an exploration of the visual culture of scientific demonstrations in this period see Stafford's *Artful Science* (1999).
4 Opinions differ as to their identities. Craske (2022) advances reasons for the man being Peter Burdett, who features in other paintings by Wright, including the *Lecture* ... on the Orrery. However, it is also possible that they are a local couple, Thomas Coltman and his future wife Mary (see Daniels n.d.), later painted by Wright. There is a certain physical resemblance between Coltman and Burdett.
5 The Mediaeval system of dual principles of sulphur, associated with fire, and mercury, associated with liquidity, was expounded by the Arabian alchemist Abu-Musa-Jābir-ibn-Haiȳan (c721–815) (Latinized as "Geber"). Salt, representing solidity, was later added to form a trinity, a system adopted by Paracelsus (c1493–1541) (see Dampier 1948 and Roob 1997).
6 The term 'Industrial Revolution' is thought to have been first used by Engels in 1844 – see Bernal (1969, 520).
7 The preparatory drawing of *The Alchemist* by Pieter Brueghel the Elder (1558) is held at the Kupferstichkabinett Berlin – a number of copies after the work exist.
8 See the Wellcome Library collection https://wellcomecollection.org/works/qrdfgfy8
9 A detailed account of the process used by Macquer is given in Powers, J. (2020). *Searching for the Philosopher's Stone: The Story behind Joseph Wright's Alchymist*. Derby: Quandary.

Bibliography

'Alchemy' (1984) *The Shorter Oxford English Dictionary*. Oxford: Clarendon Press.

Arendt, H. (1998) *The Human Condition*. Chicago, IL: University of Chicago Press.

Arendt, H. (2016) *On Revolution*. London: Faber & Faber.

Bacon, F. (2004) The advancement of learning (First published 1605) http://www.gutenberg.org/ebooks/5500 (Accessed 1 September 2022).

Bamford, L. and Wallis, J. (2017) *Joseph Wright of Derby: An Introduction to His Life and Work through the Collection at Derby Museums*. Derby: Derby Museums.

Bernal, J. D. (1969) *Science in History* (4 Volumes). London: Penguin Books.

Burke, E. (2004) *A Philosophical Enquiry into the Sublime and Beautiful*. London: Penguin Books.

Burtt, E. A. (1967) *The Metaphysical Foundations of Modern Physical Science*. London: Routledge.

Craske, M. (2022) *Joseph Wright of Derby: Painter of Darkness*. London: Paul Mellon Centre for Studies in British Art.

Dampier, W. C. (1948) *A History of Science: and Its Relations with Philosophy & Religion*. Cambridge: Cambridge University Press.
Daniels, S. (1999) *Joseph Wright*. London: Tate Gallery.
Donald, D. (2007) *Picturing Animals in Britain: c1750–1850*. New Haven, CT: Yale University Press.
Egerton, J. (1990) *Wright of Derby*. London: Tate Gallery.
Ginsborg, H. (2022) 'Kant's Aesthetics and Teleology', in E. N. Zalta (ed.) *The Stanford Encyclopedia of Philosophy* (Fall 2022 Edition), forthcoming, https://plato.stanford.edu/archives/fall2022/entries/kant-aesthetics/
Gray, J. (2002) *Straw Dogs: Thoughts on Humans and Other Animals*. London: Granta.
Hill, C. R. (1975) 'The Iconography of the Laboratory', *Ambix*, 22(2), 102–110.
Holme, A. (2014) 'Alchemy, Image and Text: The Waning of Alchemy and the Decline of Visual Discourse in the Late Renaissance', *Journal of Illustration*, 1(2), 189–209, https://doi.org/10.1386/jill.1.2.189_1
Klossowski de Rola, S. (1973) *Alchemy: The Secret Art*. London: Thames and Hudson.
Klossowski de Rola, S. (1997) *The Golden Game: Alchemical Engravings of the Seventeenth Century*. London: Thames and Hudson.
Linden, S. J. (1974) 'Francis Bacon and Alchemy: The Reformation of Vulcan', *Journal of the History of Ideas*, 35(4), 547–560.
Linden, S. J. (2003) *The Alchemy Reader: From Hermes Trismegistus to Isaac Newton*. Cambridge: Cambridge University Press.
Locke, J. (2004) *An Essay Concerning Human Understanding*. London: Penguin.
Nicolson, B. (1968) *Joseph Wright of Derby: Painter of Light* (2 Volumes). London: Paul Mellon Foundation for British Art.
Paulson, R. (1975) *Emblem and Expression: Meaning in English Art of the Eighteenth Century*. Cambridge, MA: Harvard University Press.
Powers, J. (2017) *Experimentising the Bird in the Air Pump: Joseph Wright's Most Dramatic and Disturbing Painting*. Derby: Quandry.
Powers, J. (2020) *Searching for the Philosopher's Stone: The Story behind Joseph Wright's Alchymist*. Derby: Quandry.
Roob, A. (1997) *The Hermetic Museum: Alchemy & Mysticism*. Köln: Taschen.
Schmidgen, H. (2021). The Laboratory. *Encyclopedia of the History of Science*. https://doi.org/10.34758/sz06-t975
Solkin, D. H. (2003) 'Joseph Wright of Derby and the Sublime Art of Labor', *Representations*, 83, 167–194.
Stafford, B. M. (1999) *Artful Science: Enlightenment Entertainment and the Eclipse of Visual Education*. Cambridge, MA: MIT Press.
Todd, R. (1946) *Tracks in the Snow: Studies in English Science and Art*. London: Grey Walls Press.
Vertesi, J. (2007) *Light and enlightenment in Joseph Wright of Derby's The Alchymist*. PhD Thesis. http://www.geocities.com/jvertesi/wright/ (Accessed 9 February 2009 – no longer available. Archived copy available at: https://web.archive.org/web/20121004004754/http://www.reocities.com/jvertesi/wright/#_ftn11
Von Amelunxen, H. (2006) '*Afterword*', in V. Flusser (ed.) *Towards a Philosophy of Photography*. London: Reaktion Books, pp. 86–94.

5 Towards a New Aesthetics of Science

Aesthetic Cultures and the Processes and Objects of Regard

Adrian Currie

5.1 Introduction

Many philosophers interested in the role of aesthetic judgement and sensibility in science agree that science is, in some sense, ineliminably aesthetic. That is, if you want a philosophical story about scientific knowledge, you'll also need one about scientific aesthetics. There is much space for interesting disagreement here: we might disagree about how to characterise the aesthetic and the epistemic, and disagree on how they interact; we might disagree regarding the sense in which the aesthetic is ineliminable, and we might disagree about the role of aesthetic sensibility or judgement in shaping scientific practice.

In this chapter, my primary aim is to map out some of these possible disagreements and to provide some suggestions regarding how they might be navigated. I'll argue (1) a useful way of conceptualising positions on the aesthetic in science falls along two dimensions, one of which concerns whether the aesthetic is in service of the epistemic, or vice-versa, the other concerning the universality, stability, and source of aesthetic sensibility in science; and I'll argue (2) in deciding between views within that space we should consider the objects and processes of aesthetic regard, as opposed to starting from phenomenology or aesthetic concepts. I hope that the approach is useful for getting to grips on what we philosophers, who think the aesthetic is crucial for science, should be getting on with.

I'll start Section 5.2 by discussing what we might mean when we say we're interested in the aesthetic in science. Instead of a strict analysis, I'll point to a bundle of properties and practices that plausibly fit under the umbrella. In Section 5.3, I'll provide an analysis of what I'll call an *aesthetic culture*, that is, the aspects of a scientific discipline or sub-discipline that utilise shared aesthetic sensibilities and judgements. In Section 5.4, I'll sketch a philosophical space by (opportunistically!) situating four philosophical views along two dimensions of disagreement. In Section 5.5, I'll sign off with a speculative argument towards focusing philosophical attention to what I'll call the "objects and processes" of aesthetic sensibility.

DOI: 10.4324/9781003263920-6

This is a chapter in a book about the experimental in science, but I'll not talk directly about "experiments" *per se*. However, this chapter is certainly about *experience* in science, and insofar as a focus on the experimental is a corrective to traditional philosophical interest in scientific theories and their aesthetic properties, I take myself and the other authors here to be fellow travellers.

5.2 Where's the Aesthetics in All This?

Borealopelta markmitchelli was a species of nodosaur—heavily armoured, quadrupedal dinosaurs not unlike the iconic ankylosaur—known from a single specimen dated to something around 110 million years ago (see Figure 5.1). That specimen, *B. markmitchelli*'s holotype, really is something special. The animal was preserved under extremely specific circumstances (involving it being swept out to sea), leaving its front half preserved in three-dimensions looking, once prepared, more like a statue of the living animal than anything else. Palaeontologists analysing the fossil have used it as an inroad to the adaptive colouration and predator-prey relations in such animals (Brown et al. 2017),

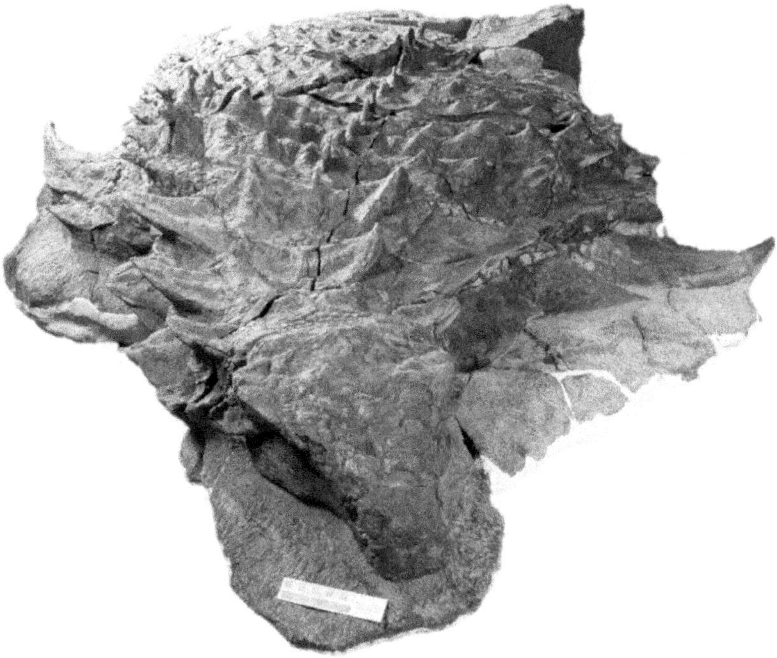

Figure 5.1 Prepared fossil of *Borealopelta markmitchelli* (detail of Figure 1, Brown et al. 2017, creative commons).

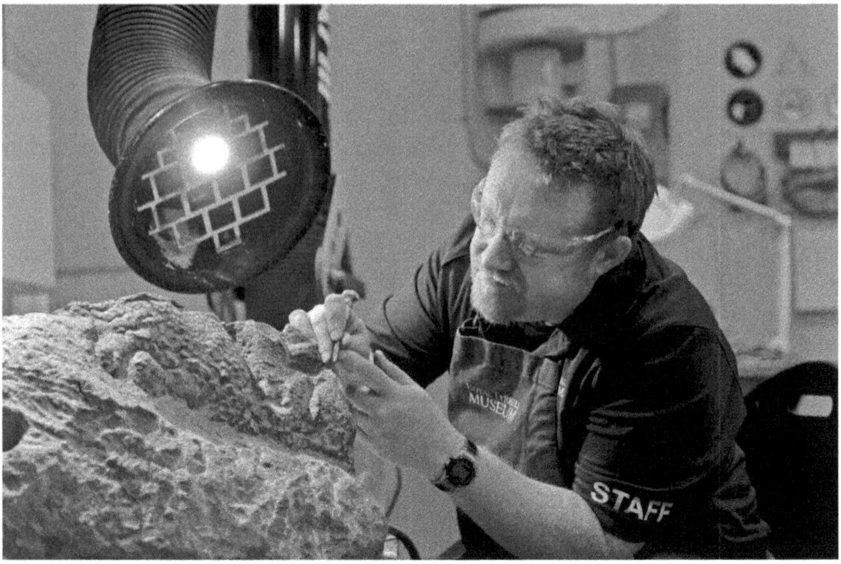

Figure 5.2 Mark Mitchell at Work on *Borealopelta markmitchelli.* Photo by Robert Clark (©) Robert Clark.

speculation fascinating in and of itself, although here we're not interested in scientific analysis, but a prior process: the fossil's preparation.

B. markmitchelli was painstakingly prepared by its namesake, Mark Mitchell. Over six years, Mitchell carefully extracted the specimen from surrounding rock, a process requiring craft, artistry and care. 'The specific epithet *markmitchelli* honors Mark Mitchell for his more than 7,000 hours of patient and skilled preparation of the holotype' (2514, see Figure 5.2). As Caitlin Wylie has shown, fossil preparators utilise aims and skills different from those of palaeontologists (Wylie 2009, 2015, 2018, 2019, 2021). She catalogues how fossil preparators emphasise their artistry in order to differentiate themselves and assert power in palaeontology labs.

> By drawing parallels with art – such as likening preparing fossils to carving sculptures – and by emphasizing creativity in solving problems, preparators present their work and their role as skillful, individualized, and irreplaceable ... in paleontology laboratories, low-status and informally trained technicians have significant control over choosing techniques. As a result, each preparator's artistic skill, aesthetic judgment, and creative problem-solving affect prepared specimens' appearance and thus scientific interpretation.
>
> (Wylie 2018, 52)

Fossil preparators can be counted as amongst the 'invisible' in science (Shapin 1989; Oreskes 1996; Anscomb 2021): rarely mentioned in publications nor presented to the public (although some museums, such as the Royal Tyrell where Mitchell works, do make their fossil preparation lab visible to the public—and Mitchell's epithet is a telling exception). However, the work of the preparator is crucial for palaeontological science and, I think, a great case for my sketch of the aesthetics of science. Moving forwards, don't be seduced by the beauty of Mitchell's final product, the *B. markmitchelli* specimen: we'll be more interested in how aesthetics features into the process of its preparation.

When we talk about an *aesthetics of science*, what kind of thing do we have in mind? Let's start by distinguishing between what I've called *high-falutin* and *low-falutin* conceptions of the aesthetic (Currie forthcoming). The former draws from a Kantian tradition, taking aesthetic value to be conceptually and experientially disconnected from our particular beliefs, desires, and cognitive faculties. As Alexander Wragge-Morley (2020) puts it,

> [For Kant] judgments of beauty depended on mental processes distinct from those at stake in the exercise of rational thought. The special form of pleasure arising from the encounter with beautiful objects, he argued, was incompatible with the cognitive activity at stake in the determination of concepts
>
> (4)

Following this tradition, our poor minds are unable to appreciate the aesthetic when 'forced to follow the logical procedures required for working out moral or scientific truths' (Ibid, 4). Rationality and aesthetic judgement are mutually exclusive. Aesthetics is crucially disinterested, and aesthetic appreciation is of "pure form". High-falutin' accounts, due to their disconnect with day-to-day human experience and the exercise of rational faculties, are of limited use in understanding the aesthetic sensibilities and judgements in as messy and human an activity as science. We might perhaps enjoy the formal properties of *B. markmitchelli's* holotype—its clean lines and pleasing geometry—but I doubt this will tell us much about Mark Mitchell's aesthetic practices. Understanding that requires pulling down the barriers between the aesthetic and the cognitive.

Recent work in the aesthetics of science harkens to an earlier conception of the aesthetic, one which Wragge-Morley associates with Alexander Gottlieb Baumgarten: 'Defined by Baumgarten as the "science of sensitive cognition", that field was concerned not merely with the experience of art but more broadly with sensory perception and its links with the other faculties of the mind' (6). Such *low-falutin'* accounts take the aesthetic to be continuous with other aspects of experience and to be (at the very least) in a rich dialogue with the beliefs, desires, and aims that come with rational action. To an extent, this approach to aesthetics sacrifices analytic specificity for usefulness.[1] Perhaps the closest I can manage definition-wise is that low-falutin' approaches are concerned with how *valenced, affective experience—and the judgements that arise from those qualities of experience—connect with human activity*. Instead of attempting anything more carefully, I'll sketch a set of approaches and concerns that have animated recent interest in the aesthetics of science. In the following section, I'll call these the "aesthetic sensibilities" that delineate an aesthetic culture.

Mark Mitchell and other preparators' practices bring *embodied, situated skills* to the forefront of our thinking about the aesthetics of science. Fossil preparation involves intimate interaction with fossil and rock matrix, often delicate and opaque. Wylie emphasises the opportunistic, methodologically diverse aspects of preparator's work, a diversity reflected in their research environments.

> ... many fossil labs look like cluttered workshops, with assorted surprising objects including power tools, delicate artists' brushes, former yogurt containers filled with freshly mixed plaster or liquid adhesive, unidentifiable metal pieces carefully bent into precise but mysterious shapes, and fossils cradled on sandbags hand-sewn from old jeans.
>
> (2021, 102)

Much of science involves the exercise of highly trained, experiential know-how situated within highly specialised environments, and this has a clear aesthetic element (Wragge-Morley 2020; Currie 2022, Roberts et al. in prep). As Mitchell works with the fossil, he makes judgements about which materials and techniques to use (Wylie recounts a near-obsession amongst preparators with forms of glue and other adhesives), about which parts of the specimen count as fossil and which counts as matrix, and about which risks are worth taking in the delicate process. As Wylie emphasises, these judgements involve affective experience and aesthetic sensibilities. Further, Mitchell exercises forms of agency as he works with the fossil, both in explicit judgement and the expression of expertise. Indeed, varieties of scientific circumstances generate varieties of affordances towards scientific exercise of agency (Pickering 2012; Murphy et al. under review).

The aesthetic judgements of Mitchell and the palaeontologists, volunteers, and preparators he works with do not stop at the relationship between fossil and matrix, they more generally concern the products of science, their objects, and their processes (Ivanova 2017, this volume). Regarding products, fossil preparation involves the creation of what Derek Turner has called *aesthetic things*. As he says, '[f]ossils are both "technical objects" of paleontological research – part of the paleontologists' toolkit – and aesthetic things of paleontological science' (Turner

2019, 54). That is, the outputs of science are not simply the objects of epistemic regard, but also count as objects of aesthetic regard. Mitchell prepares his fossil both for scientific work and for museum display, bringing together the epistemic, aesthetic, pedagogical, and economic. In doing so, he exercises not only physical skills but also imaginative faculties (Todd 2020). As Nicolas Rassmusen—in a discussion nicely dovetailing with mine—claims of various apparently epistemic decisions regarding mesosome experiments: 'Aesthetic preference influences beliefs at every stage, from the choice of areas of the specimen to photograph, to the presentations of micrographs in journals' (260).

Low-falutin' aesthetic sensibilities apply to scientific products as well: judgements must be made concerning significance (Ivanova 2022) and surprise (Currie 2018, French and Murphy forthcoming), as well as the satisficing (Gopnik 1998) nature of explanation, hypotheses, and evidence. Indeed, our aesthetic judgements concerning the objects of science have an epistemically motivating role: they demand enquiry regarding the source of the aesthetic value (Currie and Turner forthcoming). For instance, *B. markmitchelli's* striking features generate questions about both how it became so well preserved, and after the form and function of its morphology. Turner argues that identifying the match between form and function so crucial to functional morphology is itself an aesthetic judgement (Turner 2019, see also Parsons and Carlson 2008). Similarly, regarding scientific processes, both Parsons and Rueger (2000) and Ivanova (2021; 2022) emphasise the crucial match between form and function in the design of experiments; we might be impressed too by the deft use of the right adhesives and tools in Mitchell's work (see also Murphy 2020 on form and function in thought experiments).

So, scientific practice is awash with (low-falutin') aesthetic value, sensibilities and judgement: it can be surprising, deft, profound (Murphy this volume), fluent (Novaes 2019), sublime (Parsons, this volume), and well-fitted (Todd 2018). What I think is ultimately important about the aesthetic in science is its ability to open discussion of the interaction between the experience and agency of scientists (and scientific communities) and their investigative targets and tools in the production of scientific knowledge, and thereby taking seriously a set of values, judgements, and sensibilities that are often ignored in philosophical accounts. It further provides inroads for transport between the philosophy of science and aesthetics (e.g., Murphy and Anscombe's contributions to this volume).

As a quick aside, plausibly much philosophical and scientific squeamishness about the role of aesthetics in science is due to aesthetics' perceived pernicious subjectivity. If beauty and the eye of the beholder stand in a conditional, necessary relationship, then admitting aesthetic judgement into scientific judgement surely undermines the epistemic goals and prowess that science is supposed to possess. I find this unmoving, at base because it makes unlicensed assumptions about the subjectivity of aesthetic judgement: that it is somehow unmoored from, and undermining of, scientific judgement. But I see no good reason to think this. If epistemology is shaped in dialogue with aesthetic sense, then presumably aesthetics is shaped by knowledge. This kind of insight should be familiar to those versed in feminist approaches to value in science. For a single example, take Elizabeth Anderson's discussion of the relationship between moral judgement and scientific judgement:

> … science is value-free if and only if values are science-free … the arguments for neutrality depend upon crude models of how value judgments work
>
> (Anderson 2004, 7)

So long as there is a two-way dialogue between aesthetic value and science, then we shouldn't expect our epistemic judgements to be neutral pertaining to the aesthetic, but nor should we consider this a problem, so long as the aesthetic judgements do not alone determine the outcome of our scientific pursuits, and so long as scientific work in some regards speaks to the aesthetic. Unless aesthetic sensibilities deserve some kind of special pleading vis-à-vis their relationship with scientific knowledge, the various strategies developed in the 'values in science' literature are easy fodder for those of us interested in the aesthetic (see also Ivanova forthcoming).

5.3 Aesthetic Cultures

Particularly in the wake of Kuhn's influence, historians, philosophers, and sociologists of science have paid much attention to professionalisation (Currie 2019; Stanford 2019), pedagogy (Biesta and Miedema 1990), and sociality in scientific practice. If we think the aesthetic plays a role in shaping science, then, we should expect it to be prominent in how these factors shape scientific experience. Thus far, the aesthetic plays little role in these discussions (other than in McAllister's work, see below). My strategy in this section is to sketch some work on scientific *cultural identities*, and then provide an aesthetic spin on these. Specifically, I'll identify an aspect of cultural identity—the *aesthetic culture*—in reference to Section 5.2's examples of low-falutin' aesthetics in science. So, what is "cultural identity"?

Is a "cultural identity" the same as a paradigm? The term "scientific paradigm" is no synonym for "theoretical framework"—it is not best thought of a set of abstract equations, say—rather, a paradigm is a rather nebulous set of joint, often tacit, knowledge, and perspectives shared through a scientific community, that plays a crucial role in shaping what counts as an important question, the relevant approaches to it, and what counts as a satisfactory answer (Kuhn 1970). If you want, scientific cultural identity is a way of considering paradigms, although it is perhaps somewhat broader. Similar could be said of other frameworks developed for thinking about the social structure of science (e.g., Love 2008; Leonelli and Ankeny 2015; Ankeny and Leonelli 2016). So, I'm not concerned whether "paradigm" and "disciplinary culture" are co-extensive. I'll follow the archaeologist Stephanie Moser's discussion:

> [historians of science] ... have shown how professionalization involves the delineation of appropriate modes of practice and behavior and the creation of a sense of a common sense of identity. This sense of community identity incorporates notions of what it means to participate in the life of the discipline, and the formation of a professional or disciplinary culture is crucial in shaping the type of research that is done.
>
> (Moser 2007, 240)

Moser discusses the relationship between fieldwork and gender in archaeology, and I'll use this as a guide. Similar, I imagine, could be said drawing on the anthropology of experimentation (Latour and Woolgar 1979), but Moser's work suits me fine.

Moser (2007) argues that experiences of fieldwork play a crucial role in setting the cultural identity of archaeology, drawing on her own experience, documentary analysis, and discussions with other practitioners. She identifies a set of factors which set disciplinary culture, particularly the gendered aspects within archaeology (see also Cobb and Croucher 2016). First, various aspects of ideology are communicated via species of myth-telling. Consider, for instance, how Moser's informants discuss fieldwork.

> [informants emphasized] ... the qualities of landscape and also the solitude, the closeness to nature, and the engagement with previously unseen cultural relics. Essentially, practitioners talked about their fieldwork experiences in a way that was different from the manner in which they talked about other aspects of their careers. They romanticised fieldwork, conveying the idea that it evoked a sense of adventure and discovery, or freedom, of going where no one had been, of challenge, of survival, and of experiencing the landscape
>
> (Moser 2007, 249)

Fieldwork certainly evokes aesthetic responses, and appreciation of these features forms part of the cultural identity archaeologists adopt. I think it likely as well that these sensibilities make a difference to archaeological practice: where fieldwork happens, how it is conducted, and so forth.

Second, epistemic (and other) authority is granted on the basis of various hurdles being overcome. The rough-and-tough (and, as Moser emphasises, masculinised) world of the archaeological field is taken as a kind of rite of passage, surviving or thriving within which is required for one to count as an archaeologist. Similar might be imagined of the various trials of mundane laboratory work, or the experience of completing a PhD, or delivering a research seminar to a hostile philosophical audience. Such experiences seem to both encode what matters to that discipline's culture—what its values and self-conception are—and to gate-keep entry.

Third, such discipline-specific experiences serve to form a sense of identity within that community, and also a form of autonomy. As Moser discusses, fieldwork acts as a kind of "epistemic anchor" in archaeology; what makes archaeology *archaeology* is the confrontation between archaeologist, specimens, and assemblages in the field. This idea is carried into post-processualist conceptions, most explicitly Ian Hodder's claim that interpretation begins 'at the trowel's edge' (Hodder 2003).

Returning to fossil preparation, Wylie's accounts of how preparators construct and utilise their disciplinary culture fit Moser's discussion of cultural identity well:

> Choosing artistry and creativity as group-defining concepts portrays preparators as free-spirited and independent makers. As such, they view their field as the skillful, hands-on assessment and alteration of objects' characteristics, rooted in nonstandard ideas of good practices and in the contexts of researchers' diverse purposes for prepared specimens. In their domain, most preparators have primary control, without protocols or methodological instructions from researchers. Furthermore, preparators themselves manage who enters their field without researchers' intervention, through informal local training of novices. It is that independence, I argue, that drives preparators' senses of personal identity and shared community.
>
> (Wylie 2015, 52)

The non-uniform, idiosyncratic methodological culture of fossil preparators serves to both shape a disciplinary culture, police entrance to it, and establish autonomy and control for the preparators.

So, a crucial part of scientific communities is the *disciplinary culture*, which at base we can understand as the shared sense of identity that forms the dominant ideology, identifies the epistemic anchor, and governs autonomy and authority within the discipline. Disciplinary culture emerges from and is maintained via, pedagogy, professionalisation and socialisation. Part of disciplinary culture, too, is the mundane day-to-day skills and practices underwriting scientific work. Pedagogy within archaeology, geology, and other field sciences are often situated around field schools where the procedural uptake of skills—as well as socialisation—comes together (Munge et al. 2018).

Reflecting on Section 5.2, we can see that much disciplinary culture is aesthetic. No doubt, the romanticism of Moser's archaeologists is clearly aesthetic, but further, much of the low-falutin' aesthetic skill, perspectives, sensibilities, and judgements of a discipline will form part of its culture. The aesthetic judgements of many archaeologists were likely shaped in the field: along with the myth-making and gate-keeping, passed along too is the embodied skill, forms of agency, appreciation of aesthetic things, judgements of aptness, significance, and so on. Judgements about, say, an excavation's cleanliness, order or beauty, are rooted in embodied experiences in the field. And much the same could be said of work in the fossil preparation lab, as novice preparators are shown the ropes by old hands—ushered through the gated community of fossil preparation—taking on the aesthetic elements of the culture as well. And presumably various—potentially highly diverse—rites of passage exist across the sciences. As such, let's coin a term: *aesthetic culture*, and give it something of a definition:

> A set of individuals that (1) share cultural identity and (2) share relevant low-falutin' aesthetic sensibilities, may be considered to share an *aesthetic culture*.

Let's unpack this. By "aesthetic sensibilities", I mean the package of skills, valenced affect, judgement, norms, and suchlike, as discussed in Section 5.2. I take it that sharing cultural identity requires continuity in the kinds of activities and processes that Moser discusses. A history of fieldwork provisions archaeologists with a shared history that shapes their practices. By "shared sensibilities", I mean something approximate: there's unlikely to be an exact match across such a messy set, but insofar as we can make sense of what it is to "partake in a paradigm" I assume we can get a similar grip on what it is to share aesthetic sensibilities. Scientists sharing aesthetic sensibilities will make the same (or relevantly similar) judgements about significance, satisfaction, and so forth. As is hopefully clear, the scope of aesthetic culture is an open question: they could be discipline-wide but could be significantly more local. That is, aesthetic cultures, or aspects of them, could well apply to traditional disciplinary boundaries, reach beyond them, or find themselves restricted to particular subdisciplines, labs, or (at the most extreme) particular scientists in particular contexts. Such questions of scope will be central to one point of disagreement I'll outline below.

Crucially, different aesthetic cultures might overlap in various spaces of scientific work—perhaps to that work's benefit or detriment. As we've seen, Wylie emphasises the differences between preparator and palaeontologist, differences that track their aesthetic expertise. Palaeontological aesthetic cultures involve particular ways of thinking about and interacting with fossils: their analysis and interpretation as data for the past. The aesthetic culture of fossil preparators differs: fossils are targets for their artistic skill, things to be carved out of their matrices. Understanding the harmonies—and clashes—of these two aesthetic cultures could be understood as a major focus of Wylie's work. You might also wonder after the relationship between an individual's aesthetic sensibility and those of the aesthetic cultures they partake in. There's far too much to wonder after to cover here, but open questions include whether individuals are in a single culture or many cultures and whether aesthetic sensibility is based within agents or distributed across groups.

5.4 Four Views, Two Dimensions

I've sketched a set of low-falutin' aesthetic elements in science and connected these to scientific communities via the notion of an aesthetic culture. With these tabled, I can now turn to a set of philosophical views regarding the relationship between the aesthetic and the epistemic in science. Although the philosophers I'll discuss do not explicitly talk about aesthetic cultures, I think they are fruitfully taken as such. And I think these differences are usefully understood as answering two questions. To avoid burying the lead, these are:

> Q1: What drives aesthetic cultures? (epistemic-led or aesthetic-led)?

That is, is the aesthetic community largely determined by epistemic or aesthetic concerns (or something else, or a combination)?[2]

> Q2: How malleable are aesthetic cultures?

That is, what determines aesthetic sensibilities? Are they innate and hard to shift, are they determined by very general epistemic principles and hard to shift, or are they highly malleable and local (and if so, what factors determine them)?

James McAllister has developed an explicitly Kuhnian story about the relationship between science and aesthetics (1996a; 1996b). He is particularly interested in how the aesthetic aspects of a paradigm, especially shared aesthetic judgements about the good-making properties of theories (their elegance, simplicity, etc. ...) might influence and interact with paradigm shifts. On McAllister's view, the "aesthetic canon" of a scientific discipline is stable and unified across what Kuhn would call *normal science*—the day-to-day problem-solving that constitutes the majority of scientific work—but the canon itself may undergo paradigm-like revolutions. So, the aesthetic canon may remain stable across changes in the scientific bedrock, thus providing continuity and constraint across scientific change, but it can also produce ruptures driving science's shape in surprising directions.

McAllister's story about aesthetic canons adopts a decidedly un-Kuhnian rationalist spin. He argues the rationale for the aesthetic canon ultimately rests on an optimistic meta-induction across a history of past success. Because, say, simple theories have proven fruitful in the past, our aesthetic preferences for simplicity are likely to be knowledge-tracking. There is thus an epistemic grounding for the role of aesthetic sensibility in science. McAllister doesn't necessarily think that every (or any) individual scientist makes this judgement, but rather that previous success explains the acceptance and dissemination of those aesthetic values. As he puts it,

> … while a community's aesthetic canon changes with time, the aesthetic preferences of scientists at any one time do not diverge very strongly: there is wide agreement about the aesthetic properties that theories should possess. This is because, far from being conceived at whim, scientists' aesthetic preferences are formed in a communitywide induction over the empirical performance of past theories.
>
> (McAllister 1996a, 137–138)

Milena Ivanova (2020) interprets the proximate mechanism by which the aesthetic is passed on in McAlister's view as *exposure theory*, a (likely false, or at least counterexample-ridden) psychological theory which, approximately, has it that continual exposure to a set of properties over a relevant developmental window will produce positively valenced attitudes towards those properties. Over years of scientific training, scientists are exposed to and encouraged to interact with, multiple examples of simple theories, and so "simplicity" will begin to be expected—aesthetically and normatively expected—of good theories. By contrast to McAllister's view, Ivanova argues that aesthetic sensibilities are significantly more resistant to the influence of pedagogy. Ivanova casts doubt on McAllister's meta-induction, providing examples of aesthetic preferences that have stuck around despite an apparent history of failure. She explains such phenomena by appealing to aesthetic preferences that are 'deeply ingrained in human thinking' (Ivanova 2020, 89, see also Ivanova 2017). Because of this, it is not past epistemic success that shapes scientific aesthetic preferences, but our ingrained aesthetic preferences that shape our epistemic tools:

> [aesthetic properties] are properties of our theories only because we decided to construct them in such a way, and we decide to construct them this way because it is most convenient for us to operate with theories that satisfy our aesthetic and intellectual requirements.
>
> (Ivanova 2020, 90)

Ivanova's differences with McAllister provide a useful contrast given the two organising questions I mentioned above. Where McAllister thinks that (ultimately) epistemic preferences are in the driver's seat, Ivanova more emphasises the influence of aesthetic preference (although whether she would ultimately commit to the view I'm drawing from her here is an open question). This comes with differing views regarding the malleability of aesthetic cultures. For McAllister these follow a punctuated pattern where periods of stability and continuity are occasionally disrupted by revolutionary change. And perhaps for Ivanova, although there might be local differences in how they are treated in various aesthetic cultures, given that the same ingrained aesthetic values underwrite them, we shouldn't expect ruptures or large discontinuities across contexts. For McAllister, aesthetic cultures are somewhat dynamic through a revolutionary change, but fairly stable otherwise. For Ivanova, aesthetic culture is likely broad, with much agreement between disciplines (or at least within a discipline). This difference is partly due to their differing views on who is in the driving seat.

Neither Ivanova nor McAllister is particularly explicit about the conceptual relationship between aesthetic and epistemic concepts, a discussion on which Derek Turner focuses (Turner 2019). For Turner, there are three views we might take. First, we might think the relationship is *accidental*, which is to say, the main business of science is epistemic, but it nonetheless has aesthetic payoffs—payoffs that don't make a significant difference to the practice of science.[3] Second, according to an *intertwined* view, we might think that aesthetic and epistemic judgements are separate but mutually supporting. According to a third—and more extreme—view, the two are *blurred*, that is, there isn't a conceptual difference between them: epistemic judgement just is an aesthetic judgement, and *vice-versa* (see also Rasmussen 1993).

Turner weakly prefers blurring over intertwining, although admits he has no knock-down argument. His preference is, I think, partly explained by his account of aesthetic things. As fossils are technical—epistemic—objects, as well as aesthetic things, 'Understanding this dual role that fossils play is crucial for understanding the practice of paleontology' (54). In addition to being a technical object deployed by palaeontologists, the holotype of *B. markmitchelli* is at the very same time an aesthetic thing, the target of aesthetic judgement, and an item meant to be observed and examined both in scientific and pedagogical contexts. As Turner has pointed out to me, the specimen is mounted in a metal armature, built especially to not only hold the fossil but to imply the shape of the living animal's back half. As he put it 'I'm not sure you can totally separate the prepped fossil from the armature, and the composite of the two is, just straightforwardly, a work of art (I think)' (Turner, personal communication). In prepared fossils, the aesthetic and the epistemic seem to blur. Turner also emphasises the expertise behind aesthetic skills in scientific practices such as fieldwork and fossil preparation, emphasising the learnt aspects. His view, then, plausibly agrees with McAllister regarding the flexibility of aesthetic culture, but sits in-between Ivanova and McAllister regarding the relationship between the aesthetic and the epistemic: if there is no real distinction between the two, one or the other cannot be in the driver's seat.

My own view agrees with Turner and McAllister insofar as I think the aesthetic is dynamic but allows much more idiosyncrasy than McAllister regarding aesthetic cultures (Currie forthcoming). I argued that scientific practice employs *attuned* aesthetic judgements. In being adopted into an aesthetic culture through pedagogical and social practices, a scientist's aesthetic judgement is co-opted towards scientific purposes. Mitchell's basic aesthetic sensibilities are geared towards the needs of palaeontology—and of museum curators and of course fossil preparators themselves!—though his training in fossil preparation. This makes room for idiosyncrasy depending on the specific features of that training. Although this puts epistemology in the driving seat, I depart from McAllister insofar as I think the justification of aesthetic judgements in science is grounded in the processes underwriting learnt expertise, not a history of success. For me, attunement depends critically on the specific pedagogical and scientific practices at hand, thus the justification of aesthetic judgement turns on pedagogical and scientific context, not induction over a history of success.

With this discussion on the table, we can see how myself, Ivanova, Turner, and McAllister could be situated into a two-dimensional space (Figure 5.3). In the figure, we arrange views in terms of how the two questions set above are answered.

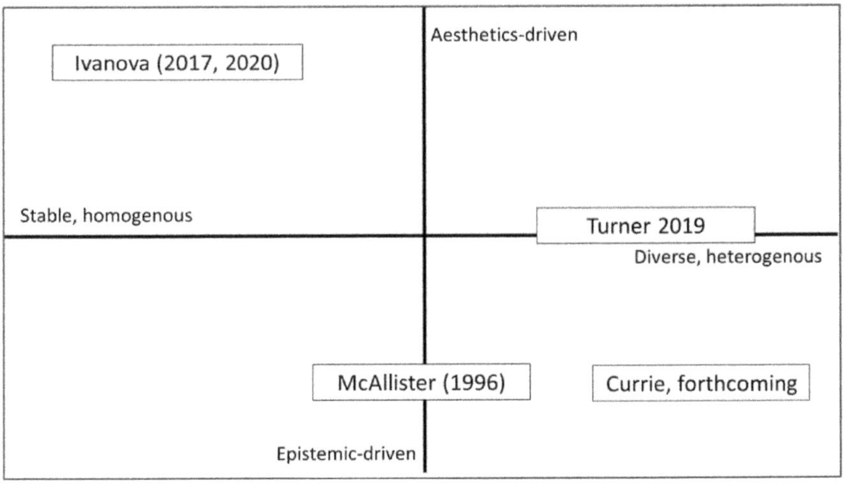

Figure 5.3 Views in the aesthetics of science. The horizontal dimension tracks the expected stability and diversity of aesthetic cultures over time and across contexts, the vertical dimension that shapes aesthetic cultures.

I have a few initial points to make regarding this organisational schema. First, note that it abstracts from *why* views are situated where they are in the space. McAllister and Ivanova see aesthetic cultures as more stable than myself and (I suspect) Turner, but for very different reasons: the former because the cultural and epistemic continuity of normal science stabilises the aesthetic canon, the latter because aesthetic judgements are deep-seated and not easily shifted over ontogeny. Second, consideration of these four philosophers leaves much space unoccupied. We might ask what views in other areas of the space would be like, and indeed this could lead us to ask further questions about which stabilising mechanisms determine positions on the malleability question, and what kinds of analyses might shift us along the epistemic vs aesthetic dimension (the top-right location, where a highly malleable aesthetic sensibility is in the driving seat, is particularly interesting).

Third, and I think, most importantly, although I have presented the views as if they are all-or-nothing affairs—as if we should commit to a single answer as to the stability of aesthetic cultures and the relationship between the aesthetic and the epistemic—I'd be astonished if a single view applied across cases. I find it extremely plausible that different aesthetic values in different scientific contexts might yield varying responses to my guiding questions. Some aesthetic values might be deep-seated and resistant to co-option, and so might end up aligning with the view I take Ivanova's discussion to imply, while others could be significantly more malleable. Others might be very clearly working to support the epistemology, while sometimes the epistemology is clearly in service of the aesthetics. The question is not, I think, which part of the space we should ultimately and finally occupy. Rather, the question is *in what contexts* should we expect the aesthetic and the epistemic to come together in particular locations in the space. Above, we saw how varied the aesthetic in science is, ranging from judgements about aptness and fit, to embodied skills, to agency, to the production of aesthetic things. I see no reason to expect a single answer about the relationship between scientific aesthetics and scientific knowledge to hold across these.

Fourth, we might distinguish between where in fact various scientific contexts fall in the space, and where they ought to, given the various needs and aims those practices serve (i.e., we might ask whether the aesthetic culture is 'well-adapted' (Currie 2019). I've no space to give the pandora's box of normativity justice here, so I'll leave it slightly ajar and move on.

I am more convinced of the flexible, plastic—"attuned"—aesthetic judgements of the lower-right corner of the space, but I hope it is clear that my aim here has not been to defend that. Rather, I've tried to do some fruitful systematising. This has likely involved a fair bit of shoe-horning, and I trust the authors at hand will forgive me for this, but I also think that this way of conceiving of the space does real work. So, with the space in place, how should we navigate it? What features of aesthetic communities should we care about in reconciling these philosophical disagreements? Indicating a move in that direction is my final task.

5.5 The Objects and Processes of Aesthetic Regard

In addition to focusing on theories, most discussions of the aesthetics in science begin with aesthetic *concepts*, that is, identifying and analysing aesthetic notions—beauty, elegance, ugliness, and so on—and asking how they fit into science's apparently epistemic concerns. I suspect this approach won't do for understanding low-falutin' scientific aesthetics and will suggest a different approach.

Thus far, I've identified two loci of disagreement concerning aesthetic communities: first, their relative stability and uniformity, and second, whether these are shaped primarily by aesthetic or epistemic ends. I don't think there's a single answer to these questions—and don't here want to attempt to argue for one perspective over another—instead, I want to consider *how* said arguments should be had. That is, when we ask after those loci of disagreement, what should those disagreements turn on? Here, I'll argue for a methodological point: they should start with what I'll call *the objects and processes of aesthetic regard*.

Well, that's a mouthful. What are the "objects and processes of aesthetic regard"? Scientists make aesthetic judgements—apply their sensibilities—to particular objects, be they theoretical abstracta, data, specimens, visual aids, experimental apparatus, and so on. For 6 years, Mark Mitchell's sensibilities were trained on the specimen of *B. markmitchelli* and its rock matrix, judging which parts to focus on, which tools to use, which risks to take, and so on. These partially aesthetic judgements are made via various processes: experimentation, data-analysis, eyeballing, etc. ... In utilising aesthetic sensibility, Mitchell engaged in various processes of aesthetic judgement: observation, tactile exploration with fingers and instruments, imagining in his mind's eye. Once when I visited him working, I noticed he had a toy nodosaur handy. Mitchell explained that he used it as a kind of comparison for the animal appearing underneath the matrix. Both the toy and the specimen, and their comparison, were objects of aesthetic regard. For instance, Mitchell discovered surprising traits in *B. markmitchelli*, such as enormous spikes along its side, traits not present in the toy.

The point of this notion ("objects and processes of aesthetic regard") is to emphasise both scientific practices—what scientists do—and also the objects that scientists utilise and target. And also, naturally, their interaction. It is thus not simply about practice, nor merely about ontology.

My position is: in asking whether the shape of science turns on aesthetic or epistemic factors, and whether aesthetic sensibilities are stable, we are asking questions about the objects and processes of aesthetic regard. What are the other possible options? I can think of two potential competitors. One is *phenomenology*. Is the quality of experience across an aesthetic culture stable? Is Mitchell's experience similar to, or different from, other fossil preparators—or palaeontologists? Or any old scientist? And is it features of that experience that shapes his behaviour? Another competitor is *concepts*. Do Mitchell and others in his aesthetic culture employ the same aesthetic ideas—are there, say, the same concepts of "beauty", "elegance" at play across the group? Plausibly, which starting place we choose makes a difference to how we interpret the guiding questions from the last section, and how we pursue them. Each option no doubt has merit—I'm nothing if not a methodological pluralist—but thinking a little about their relationship reveals some of the advantages of the focus I suggest here.

Towards that, join me in an imaginative exercise.

If we engage in a thought experiment wherein we hold fixed two of our three factors (the phenomenology, the concepts, or the objects and processes) and vary a target factor, will that make a significant difference to the knowledge at hand? Let's start with phenomenology.

Imagine we hold fixed Mitchell's aesthetic concepts. He possesses the same notions of beauty, completeness, etc. ... and holds fixed the objects and processes of his aesthetic regard—the same features of the specimen are salient to him—but changes the phenomenology. Perhaps we have something akin to a spectral inversion, say: intervene such that it is like something different for Mitchell. Will this make a difference to the knowledge he produces? Should we, say, expect a different prepared fossil to result? I think our judgement should be *no*. Without changes in concepts or objects/processes, it is hard to envision a difference in phenomenology having that kind of upshot. No doubt that phenomenology is a crucial part of an aesthetic culture—and I bet there is plenty of uniformity across members—but it doesn't seem as if continuity in phenomenology should be counted as *necessary* for membership in that culture.

Imagine we hold fixed Mitchell's phenomenology and the objects and processes of his regard, but intervene on the concepts he holds. Here, it is tempting to think we should see changes in the scientific output. After all, if two scientists have differing notions of simplicity, say, and simplicity judgements explicitly shape how they do their science, then we should expect their epistemology to diverge, potentially radically. I'm not convinced, however, and for several reasons. First, our aesthetic lexicon is often impoverished. Given the shades of subtlety we have concerning actual judgements of things like simplicity, our available concepts appear rather blunt. Second, in contexts where apparently aesthetic judgements are codified, we've drifted a long way from anything explicitly aesthetic. Consider, for instance, phylogenetic parsimony, where "simplicity" means "a phylogeny with minimal branching characters, minimising the number of evolutionary events". Here, plausibly, we're simply talking about an epistemic principle grounded in general expectations of how that kind of system tends to behave (Sober 1988; Currie 2021).

Third, consider a more plausible approach, which takes aesthetic concepts to be at least partly tacit, perhaps emerging from scientific practice. Here, aesthetic concepts that scientists deploy are not transparent. They can only be identified by examining what scientists do: concept possession is encoded in deployment. This approach allows exactly the kind of flexibility and subtlety that relying on explicit concepts appears to fail on. However, this solution falls prey to a dilemma: either it collapses into consideration of the objects and processes of aesthetic regard, or it adds an unhelpful layer to our analysis. Let me explain.

On the first horn, if aesthetic concepts are rich and tacit, then this amounts to saying that they track how it is that scientists engage with the objects of their study. Mitchell surely has rich experiential knowledge of, and skill regarding, things like vertebrate anatomy, the relationship between matrix and fossil, various adhesives, and the structural properties of rock: experiential knowledge shot through with aesthetic sensibility owed to many years of practising his craft. These tacit and subtle concepts emerge from, and are understandable in terms of, the processes and objects of aesthetic regard. As such, they don't amount to a separate option.

On the second horn, we claim that there are aesthetic, tacit, conceptual resources that go beyond the objects and processes of regard and the phenomenology. What might these resources be like and how might we access them? Here, my imagination at least runs out: I think the onus is on the concept-first defender to make sense of what these resources might be and (perhaps more pressing) how we might bring them to bear in deciding between our competing positions. Further, if concepts are subtle and encoded in practice, then methodologically speaking, starting with the practice, and postulating concepts based on our understanding of the role they play in that practice, seems a much more fruitful approach than attempting to tackle the relevant concepts directly.

So, what if we hold fixed the concepts and the phenomenology but intervene in the objects and processes of aesthetic regard? Here, I think we should expect differences—that should hardly surprise you—but it is worth clarifying how. Understood naively, if this question amounted to "what if there were literally a different object?", then our answer becomes fairly direct. If Mitchell prepared a different specimen then surely he'd end up with a different product. But that is too easy. Instead, we should construe the objects and processes of aesthetic regard in terms of the various *properties* of objects that the aesthetic sensibilities are sensitive to, and how the processes at hand make those properties salient. Mitchell has prepared many more specimens than the nodosaur that bears his name, but presumably, there is continuity between his approaches: what he considers a complete or incomplete, or beautiful or ugly, or satisficing or frustrating, specimen. On this understanding, the question becomes significantly more interesting and, I think, we should still answer in the affirmative.

Had different properties of *B. markmitchelli* triggered different aesthetic judgements and processes from Mitchell, then the end result would likely differ—even if we hold the concepts and phenomenology fixed. Consider all the aesthetic judgements Mitchell had to make in preparing the specimen. Judgements about which tools to use in removing fossil from matrix, which adhesives and preservatives to apply, which features to concentrate his efforts on, when the elements of the specimen, and the specimen as a whole, was complete. As Caitlin Wylie has pointed out, data is often underdetermined by specimens. That is, what palaeontological data we get from a fossil turns crucially on how that particular specimen is prepared: '… distinguishing coveted fossilized bone from useless rock requires the tacit knowledge of expert visual judgment and manual skill to reveal the fossil by destroying the rock (i.e., the process of fossil preparation). Because of this, the same starting rock can result in a variety of different specimens' (Wylie 2019, 24). These judgements are at least in part guided by various experiential factors embedded in Mitchell's aesthetic culture.

Let's reflect on the exercise. First, a clear objection: the idealisation involved in the central thought experiment—isolating the phenomenal from the conceptual, and both from the processes and objects of regard—is unlicensed. We might say it is conceptually impossible to decouple them as I have done, or at least psychologically impossible. Second, surely, relying on a thought experiment as it does, the argumentative force of the exercise is weak. Why should we take our judgements about the products of scientific practice in wildly psychologically unlikely circumstances as guiding us about anything?

To both of these objections, I've a single response. Perhaps, despite appearances, the point of the exercise is not primarily to generate intuitive judgements supporting the argument that we should prefer the objects of regard over competitors. Instead, it is an encouragement for us to think through what might differentiate these positions. Each of them provides a kind of gloss on the two questions driving this discussion, and I think we should prefer the gloss we get from focusing on scientific objects and processes. Why?

On the gloss I prefer, questions of malleability ask whether the *same* properties of scientific objects are highlighted by aesthetic judgement and practice across contexts. The question regarding drivers now asks whether considerations of the aesthetic, or considerations of the epistemic, ultimately decide what is salient about scientific objects. These questions, it seems to me, are fruitfully tractable. They turn on what scientists do, not what they say (and I don't think how scientists talk about their work is a particularly good guide to it, at least under normal circumstances[4]). It further leads to a consideration of the materiality of scientific interpretation, and how scientific objects and processes themselves situate scientific agents and shape their work. There's a kind of sophisticated behaviourism embedded in my suggestion that I think makes for rich approaches from historians and sociologists and, indeed, philosophers prepared to grapple with practice.

So, if we're asking after the aesthetic cultures that Mark Mitchell is part of, we should attend to how the objects of aesthetic regard, most obviously the specimen, but perhaps also the toy nodosaur and other reference objects, and the various tools he makes use of, and the processes he engages in and reacts to, makes salient properties of the target object. The choices made about adhesives, say, might speak to what kinds of joints and fixes are important, and indeed, what is considered not important (fossil preparation, like fieldwork itself, is a necessarily information-destroying activity). Other fossil preparators might adopt differing styles and approaches, and these could likely speak to differing aesthetic judgements and sensibilities. In doing this, we might find that there are large-scale, stable similarities throughout aesthetic cultures, and we might find idiosyncrasies. We might find that which aspects are salient are driven by forms of pre-scientific aesthetic regard, or that a rational—epistemic—story can be told about how and why the practices play out as they do.

5.6 Conclusion

In this chapter, I've aimed to make some headway into a set of new directions for the aesthetics of science. I've sketched a broad and diverse set of properties and activities, labelling them "aesthetic" based on their involving valenced, affective experience. I've highlighted the notion of an aesthetic culture and suggested two questions for them. First, how stable and homogenous are aesthetic cultures? Second, what drives aesthetic cultures? In answering such questions, we should consider both scientific practices and the objects and entities those practices target. This project is clearly one that involves cooperation between philosophers and anthropologists, sociologists, and historians of science. Answering these questions requires—necessitates—close and careful description and analysis of scientific practices and their target objects, as well as conceptual analyses and normative arguments. If the aesthetic matters to science, then examining aesthetic cultures by attending to the objects and processes of aesthetic regard is an important project, and, I hope, an exciting prospect.

Notes

1 There might be an interesting overlap between this approach and everyday aesthetics (Saito 2007).

2 You might worry that if we think that there is no distinction to be drawn between the aesthetic and the epistemic that this question is ill-formed, see the discussion of Turner's views below.

3 It is worth noting that this characterisation brings together two logically distinct questions. The first asks whether aesthetic and epistemic concepts are logically or conceptually distinct, and the second is whether the aesthetic matters for scientific practice. I think that the aesthetic and epistemic are distinct as a matter of principle, but in *beings like us* aesthetic judgement is critical, so it does make a difference to *our* epistemic practice. I suspect Todd (2008) could be read along similar lines. In this kind of reading, accidental and intertwining views are more difficult to distinguish.

4 To head off possible misreading: I do think that careful and critical attention to what scientists say, when embedded in a thorough understanding of their practices, can be very powerful. I take Wylie's work to be an example of this, as well as Anetolii Kozlov's empirical work on scientific emotion (under review).

Bibliography

Anderson, E. (2004) 'Uses of Value Judgments in Science: A General Argument, With Lessons from a Case Study of Feminist Research on Divorce', *Hypatia*, 19(1), 1–24.

Ankeny, R. A. and Leonelli, S. (2016) 'Repertoires: A Post-Kuhnian Perspective on Scientific Change and Collaborative Research', *Studies in History and Philosophy of Science Part A*, 60, 18–28.

Anscomb, C. (2021) 'Visibility, Creativity, and Collective Working Practices in Art and Science', *European Journal for Philosophy of Science*, 11(1), 1–23.

Anscombe, C. (this volume) Beautiful Experiments in Art and Science.

Biesta, G. J. and Miedema, S. (1990). 'Pedagogy of Science: The Contribution of Pedagogy to the Philosophy of Science', *Phenomenology + Pedagogy*, 8, 118–129.

Brown, C. M., Henderson, D. M., Vinther, J., Fletcher, I., Sistiaga, A., Herrera, J. and Summons, R. E. (2017) 'An Exceptionally Preserved Three-Dimensional Armored Dinosaur Reveals Insights into Coloration and Cretaceous Predator-Prey Dynamics', *Current Biology*, 27(16), 2514–2521.

Cobb, H. and Croucher, K. (2016) 'Personal, Political, Pedagogic: Challenging the Binary Bind in Archaeological Teaching, Learning and Fieldwork', *Journal of Archaeological Method and Theory*, 23(3), 949–969.

Currie, A. (2018) 'The Argument from Surprise', *Canadian Journal of Philosophy*, 48(5), 639–661.

Currie, A. (2019) 'Existential Risk, Creativity & Well-Adapted Science', *Studies in History and Philosophy of Science Part A*, 76, 39–48.

Currie, A. (2021) *Comparative Thinking in Biology*. Cambridge: Cambridge University Press.

Currie, A. (2022) 'Speculation Made Material: Experimental Archaeology and Maker's Knowledge', *Philosophy of Science*, 89(2), 337–359.

Currie, A. (forthcoming) *Epistemic Engagement, Aesthetic Value & Scientific Practice*. BJPS.

Currie, A. and Turner, D. (forthcoming) *Creativity without Agency: Evolutionary Flair & Aesthetic Engagement*. Ergo.

French, S. and Murphy, A. (2021) *The Value of Surprise in Science*. Erkenntnis.

Gopnik, A. (1998) 'Explanation as Orgasm', *Minds and Machines*, 8(1), 101–118.

Hodder, I. (2003) 'Archaeological Reflexivity and the "Local" Voice', *Anthropological Quarterly*, 76(1), 55–69.

Ivanova, M. (2017) 'Poincaré's Aesthetics of Science', *Synthese*, 194(7), 2581–2594.

Ivanova, M. (2020). 'Beauty, Truth and Understanding', *The Aesthetics of Science*. Routledge, pp. 86–103.

Ivanova, M. (2021) 'The Aesthetics of Scientific Experiments', *Philosophy Compass*, 16(3), e12730.

Ivanova, M. (2022) What is a Beautiful Experiment? *Erkenntnis*. https://doi.org/10.1007/s10670-021-00509-3

Ivanova, M. (forthcoming). 'Scientific Progress and Aesthetic Values', in Y. Shan (ed.) *New Philosophical Perspectives on Scientific Progress*. Routledge.

Ivanova, M. (this volume) The Aesthetic Value of Experiments.

Kozlov, A. (under review) Scientific Experiments beyond Surprise and Beauty.

Kuhn, T. S. (1970) *The Structure of Scientific Revolutions* (Vol. 111). Chicago, IL: University of Chicago Press.

Latour, B. and Woolgar, S. (1979) *Laboratory Life: The Construction of Scientific Facts*. New York: Princeton University Press.

Leonelli, S. and Ankeny, R. A. (2015) 'Repertoires: How to Transform a Project into a Research Community', *BioScience*, 65(7), 701–708.

Love, A. C. (2008) 'Explaining Evolutionary Innovations and Novelties: Criteria of Explanatory Adequacy and Epistemological Prerequisites', *Philosophy of Science*, 75(5), 874–886.

McAllister, J. W. (1996a) *Beauty and Revolution in Science*. Ithaca, NY: Cornell U.P.

McAllister, J. W. (1996b). 'Scientists' Aesthetic Preferences among Theories: Conservative Factors in Revolutionary Crises', *The Elusive Synthesis: Aesthetics and Science*. Dordrecht: Springer, pp. 169–187.

Moser, S. (2007) 'On Disciplinary Culture: Archaeology as Fieldwork and Its Gendered Associations', *Journal of Archaeological Method and Theory*, 14(3), 235–263.

Munge, B., Thomas, G. and Heck, D. (2018) 'Outdoor Fieldwork in Higher Education: Learning from Multidisciplinary Experience', *Journal of Experiential Education*, 41(1), 39–53.

Murphy, A (this volume) Profundity in Scientific Experiments.

Murphy, A. (2020) 'The Aesthetic and Literary Qualities of Scientific Thought Experiments', *The Aesthetics of Science*. New York: Routledge, pp. 146–166.

Murphy, A, Currie, A and Walsh, K. (under review) Aesthetics and Agency in Experiments.

Novaes, C. D. (2019) 'The Beauty (?) of Mathematical Proofs', *Advances in Experimental Philosophy of Logic and Mathematics*, 63–94.

Oreskes, N. (1996) 'Objectivity or Heroism? On the Invisibility of Women in Science', *Osiris*, 11, 87–113.

Parsons. G. Paper from this volume: Experimental Science as Epistemic Expansion: New Work for a Theory of the Sublime (this volume).

Parsons, G. and Carlson, A. (2008) *Functional Beauty*. Oxford: OUP.

Parsons, G. G. and Rueger, A. (2000) 'The Epistemic Significance of Appreciating Experiments Aesthetically', *The British Journal of Aesthetics*, 40(4), 407–423.

Pickering, A. (2012), 'The Robustness of Science and the Dance of Agency', in L. Soler, E. Trizio, T. Nickles and W. Wimsatt (eds.) *Characterizing the*

Robustness of Science: After the Practice Turn in Philosophy of Science. Dordrecht: Springer.
Rasmussen, N. (1993) 'Facts, Artifacts, and Mesosomes: Practicing Epistemology with the Electron Microscope', *Studies in History and Philosophy of Science Part A*, 24(2), 227–265.
Rheinberger, H. J. (1997) *Toward a History of Epistemic Things: Synthesizing Proteins in the Test Tube*. New York: Stanford University Press.
Roberts, T., Walsh, K. and Currie, A. (in prep) Situated Cognition in Early Modern Experimentation: The Case of Compelled Assent.
Saito, Y. (2007) *Everyday Aesthetics*. London: Oxford University Press on Demand.
Shapin, S. (1989) 'The Invisible Technician', *American Scientist*, 77(6), 554–563.
Sober, E. (1988) *Reconstructing the Past: Parsimony, Evolution, and Inference*. New York: MIT press.
Stanford, P. K. (2019) 'Unconceived Alternatives and Conservatism in Science: The Impact of Professionalization, Peer-Review, and Big Science', *Synthese*, 196(10), 3915–3932.
Todd, C. (2018) 'Fitting Feelings and Elegant Proofs: On the Psychology of Aesthetic Evaluation in Mathematics', *Philosophia Mathematica*, 26(2), 211–233.
Todd, C. (2020). 'Imagination, Aesthetic Feelings, and Scientific Reasoning', *The Aesthetics of Science*. New York: Routledge, pp. 63–85.
Todd, C. S. (2008) 'Unmasking the Truth Beneath the Beauty. Why the Supposed Aesthetic Judgments Made in Science May Not Be Aesthetic at All', *International Studies in the Philosophy of Science*, 22, 61–79.
Turner, D. D. (2019) *Paleoaesthetics and the Practice of Paleontology*. Cambridge: Cambridge University Press.
Wragge-Morley, A. (2020) *Aesthetic Science: Representing Nature in the Royal Society of London, 1650–1720*. New York: University of Chicago Press.
Wylie, C. D. (2009). 'Preparation in action: Paleontological skill and the role of the fossil preparator', Methods in fossil preparation: Proceedings of the first annual fossil preparation and collections symposium (pp. 3–12).
Wylie, C. D. (2015) 'The artist's Piece Is Already in the Stone: Constructing Creativity in Paleontology Laboratories', *Social Studies of Science*, 45(1), 31–55.
Wylie, C. D. (2018) 'Trust in Technicians in Paleontology Laboratories', *Science, Technology, & Human Values*, 43(2), 324–348.
Wylie, C. D. (2019) 'Overcoming the Underdetermination of Specimens', *Biology & Philosophy*, 34(2), 1–18.
Wylie, C. D. (2021) *Preparing Dinosaurs: The Work Behind the Scenes*. New York: MIT Press.

6 Profound Experiments

Alice Murphy

6.1 Introduction

Accounts of aesthetic values in science have typically focused on the aesthetics of theories and mathematical proofs and take classical aesthetic concepts such as "beauty", "elegance", and "simplicity" as their starting point. More recently, there has been a move beyond this and the aesthetic dimensions of other features of scientific practice have started to be considered (Turner 2019; Bird 2020; Wragge-Morley 2020; Murphy 2020a, forthcoming; Ivanova 2021, 2022; Currie forthcoming). This chapter continues to broaden the scope of the discussion and considers the aesthetics of experiments, taken to include concrete experiments, thought experiments, and other practices in science that are discussed alongside conventional experiments. By looking to scientific experiments that are "profound", it offers a departure from a consideration of properties like simplicity and elegance. I draw on theories of profundity in philosophy of art to illuminate the features that make experiments profound. I end by showing that broadening our class of aesthetic values to include profundity puts pressure on the scope of a central view applied to aesthetics in science and mathematics that brings together the epistemic and aesthetic by associating aesthetic experiences with "feelings of fluency".

As Carroll discusses, aesthetic value can be grounded in a number of aesthetic properties. Some properties, like beauty and elegance, as often discussed in the context of theories, are productive of (sensory) pleasure. Others, "Gestalt" properties, include being unified, ordered, balanced, or chaotic. Other categories are "emotion" properties like being sombre, "taste" properties such as being kitsch, and "reaction" properties like the sublime or the comic (1999, 190). Some aesthetic properties also have an aesthetic-cognitive nature. This includes "insightfulness" or, the focus of this chapter, the "profound".[1]

"Profundity" is regarded as one of the highest values of artworks. It's a rare quality. We praise artworks not just for their formal features, but for the insights that they provide, for how they challenge us, and for how they make us see the world in a different way. Likewise, we criticise

DOI: 10.4324/9781003263920-7

artworks for being shallow or superficial. In this chapter, I draw on the discussion of profundity in philosophy of art. This, I argue, offers a useful starting point for thinking about profundity in science and, albeit with some amendments, can help capture judgements of "profound" experiments.

6.2 Theories of Profundity I: "Handling" of Profound Subject Matter

The first theory of profundity that I will focus on is developed by Dodd (2014). Dodd begins by stating that while it can be difficult to pin down in words the exact content of a profound work, artworks that are thought of as profound deal with truths concerning the human condition, morality, and our freedom (ibid., 299). But profound artworks are not just *about* such themes. Rather, they explore them in a way that 'enable us to see things that we would not have seen otherwise' (ibid.). Dodd presents the example of *King Lear* to illustrate this view: 'if we have understood the play correctly, we come to entertain the possibility that we, like Lear, are infant-like in our vulnerability, occupants of a menacing universe devoid of justice or meaning' (ibid., 300). From this, Dodd (following Kivy 1990; 1997) sets out the necessary features for an artwork (*w*) being profound:

> '(1) *w* has a profound subject matter (i.e. *w* is *about* something profound); and
>
> (2) *w* handles its profound subject matter in such a way as to elicit a deeper understanding of it or a fuller grasp of its significance (in a suitably situated and prepared appreciator)' (ibid.).

Dodd focuses his paper on these two conditions, as he is primarily concerned with a question of whether pure music can be considered profound. He notes that these conditions are necessary, but they are not jointly sufficient. For a complete analysis, a further condition (again, taken from Kivy) is added:

> (3) *w* must treat its subject matter in an exemplary way, i.e. at a 'very high level of artistic or aesthetic excellence' (Kivy 1990, 203–204).

Kivy's conceptualisation of condition (1) is narrow: profound artworks denote something profound, and they express profound propositions about their subject matter.[2] Dodd expands this condition with reference to Goodman's (1976, 1985) view that denotation is not the only form of reference. While Dodd is critical of Goodman's analysis and departs from it in certain ways, he emphasises that a way in which an artwork (including "pure music") can refer to certain properties is by "displaying" them. Which properties are displayed or made salient is dependent upon contextual features. Dodd presents the example of taking a stone into an art gallery and presenting it as an artwork. In such a case, the properties that it displays in the gallery (e.g., its particular shade of grey or its patterns) will be different than when it was outside in nature (2014, 307). More generally, Dodd argues that contextual features such as the interest we have in a particular object will shape what properties are salient to us, i.e., which ones we focus on and which we ignore (ibid., 309).

Dodd cites the third movement in Mahler's First Symphony as a paradigmatic case of profundity in music. He states how certain features of this movement, such as how it 'lurches to and fro between gravity (in the form of the funeral march) and crudity (in the form of the wild, chaotic klezmer music)', standout to us, i.e., they are displayed (2014, 309–310). Furthermore, the audience recognises how the work displays properties beyond the work itself, i.e., they point to properties in the world. In the case of this movement, 'the music comes to present experience – and particularly, Jewish experience – as emotionally and evaluatively chaotic .... If, like me, you think that life has a tendency to be like this, then the music may well strike you, as it does me, as profound' (ibid., 310).

We can also clarify condition (2). Profound artworks do not need to *say* something profound about their subject matter in the way that Kivy specifies, that is, by expressing propositions about it. Instead, Dodd states, the profundity of an artwork (whether literary, musical, or otherwise) is grounded in how it possesses 'a type of *response-dependent property*: specifically, a disposition to elicit a certain kind of merited response in the understanding appreciator'. This "merited response" in the case of profound artworks is associated with an experience of "having *seen*", i.e., gaining an understanding of or an insight into the work's subject matter. Focusing on the case of literary works, Dodd argues that profound artworks are not profound in virtue of encoding a message. Rather, they allow us to "see" something in virtue of how the author constructs the text in such a way that 'prompts the understanding reader to appreciate things—things of fundamental import—that she might otherwise have overlooked' (ibid., 317).

How does this theory of profundity apply to experiments? Extending this view gives us the argument that profound experiments are those that (1) are "about" something profound and (2) handle this subject matter in a way that brings about a deeper understanding of it or appreciation of its significance and (3) treat this profound subject matter in an exemplary way.

First, let's begin with the second condition, in order to clarify the sense of "handling" in experimentation. Theodore Arabatzis's (2008)

Routledge companion piece on experiments begins with a quote from Schrödinger:

> 'The physicist has full liberty to interfere with his object and to set the conditions of experiment at will. This empowers him to invent methods widely different from, and largely superior to, the placid observation of the astronomer'
>
> (Schrödinger 1955, 13)

As Schrödinger highlights, experiments are artificial setups. The experimental object, the object that is manipulated in the experiment, departs in significant respects from the target system, i.e., the object as it is in the natural world. Scientists design and construct their experimental set-up in such a way to give us access to some phenomenon that we would not see, or would be difficult to see, in its "raw" state in nature. But of course, the experiment is taken to tell us something about the world beyond this set-up.[3]

For Elgin, the artificial nature of experiments invites a comparison between science and art. Like Dodd, Elgin develops Goodman's idea of exemplification and claims that both works of art and scientific experiments 'underscore, display, or convey some of their own features'. In doing so, they therefore present certain features in such a way that makes them salient to us, resulting in them being 'available for examination and projection' (1993; 13). Elgin compares the Michelson-Morley experiment (discussed in Ivanova's chapter in this volume) and Jackson Pollock's painting *Number 1*. These do not express a proposition about their subject matter nor do they "state a truth": 'Neither needs to. Each makes its case effectively without saying a word' (1993; 15). The experiment *demonstrates* that the speed of light is constant and *exemplifies* the absence of the ether.

With this background on experiments in place, we can now start to think about Dodd's view on profundity in the context of profound scientific experiments. The first example that I will discuss is the double-slit experiment with single electrons. This has been voted the most beautiful experiment in physics (Crease 2002). This experiment is a version of Young's double-slit experiment that demonstrates how light interferes with itself when passing through a pair of slits.

In his lecture course, Feynman (1965) described a version of Young's experiment in which electrons are used instead of light. Feynman proposed that if single electrons were fired at a screen with a double slit, then this would produce an interference pattern. Since one electron is being fired at a time, the pattern would not be caused by many electrons colliding with each other but would rather slowly form from the accumulation of single electrons. This demonstrates the dual wave-particle nature of matter: the electrons move through both slits at once like a wave but hit the detector at the other side at a single point, behaving like a particle. Feynman stated that this could only ever be realised as a thought experiment, because, he claimed, firing one electron at a time and detecting them in the way described would be a physical impossibility (1965, vol. 3, 1.4). But the experiment has since been conducted (with single electrons first in Bologna in the 1970s) and is still being improved.

There are various ways in which we can aesthetically appreciate this experiment along the lines of Ivanova's categories as set out in Chapter 1 of this volume. For example, we could find beauty in the spectacle – we could appreciate the experimental phenomena, i.e., the interesting inference patterns produced. We could also appreciate how conducting the experiment required creativity and technical skill, given how difficult it was to perform (so difficult in fact that Feynman thought it was possible only in thought). However, another aspect of what is celebrated and the reason why the experiment is considered the most beautiful experiment is also due an interaction between its profound subject matter and the particular way in which that is captured by the experimental set-up. This is due to how the experiment is *designed* (for a discussion of experimental design in relation to beauty, see Parsons and Rueger 2000; Ivanova 2021; 2022, this volume). As seen, Dodd does not specify what profound content is, pointing instead to larger "profound" themes that artworks can explore. His focus lies on the conditions for a profound artwork, rather than profundity itself. In the case of the double-slit experiment, we have an experiment that deals with or is "about" the behaviour of matter at its most fundamental level. Hence, it has a profound subject matter.

The double-slit experiment is a profound experiment due to how this profound content is revealed. The experiment is a powerful demonstration of quantum mechanics. In Elgin's terminology, it exemplifies wave-particle duality. Furthermore, it does so in a particular manner, presenting observers with the implications of a counterintuitive and abstract theory in an immediate, perceptible way. This would never be available outside of the controlled, artificial conditions of an experiment. This is why Feynman described the experiment as getting to the 'heart' of quantum mechanics, as containing its 'only mystery' (1965, vol. 3, 1.1). As Crease argues, seeing the experimental phenomena has a "transformative" impact on the viewer, convincing them of the theory's truth. Furthermore, despite its 'revolutionary result', and we can add, its profound subject matter, the experiment is 'readily understandable' (2002, 20). For these reasons, the experiment is regarded as aesthetically valuable and one of the most, if not the most, beautiful experiment in science.

We can also think about profundity as in Dodd's view in the case of thought experiments. Physicist Sean Carroll's favourite "deep, elegant or beautiful explanation" (edge.org survey, 2012) is one of Einstein's elevator thought experiments. As Carroll outlines the example:

> 'Imagine an astronaut in a spaceship with no windows, and no other way to peer at the outside world. If the ship were far away from any stars or planets, everything inside would be in free fall, there would be no gravitational field to push them around. But put the ship in orbit around a massive object, where gravity is considerable. Everything inside will still be in free fall: because all objects are affected by gravity in the same way, no one object is pushed toward or away from any other one. Sticking just to what is observed inside the spaceship, there's no way we could detect the existence of gravity.'

Carroll states how 'Einstein, in his genius, realized the profound implication of this situation': and this was that gravity is not a force at all, but it's a feature of spacetime – associated with the curvature of spacetime. In this case, we again have a profound subject matter "revealed" to us in a particular way due to the design or the formulation of the thought experiment. We could take pleasure in the content of the thought experiment or in the formal properties alone (such as the imagery of the astronaut and the spaceship and so on). But what is relevant here, is this interaction between the significance of the content of the thought experiment and how the details of the thought experiment function as a way of bringing that content about. This content (which otherwise might be difficult to apprehend or to explore) is presented through the particular elements of the scenario in such a way that brings about an understanding of its significance.[4]

To summarise so far, Dodd (following Kivy) argues that profound artworks have to meet three conditions: (1) have a profound subject matter, (2) handle that profound subject matter in a way that brings about understanding of it and its significance, allowing us to "see" something we otherwise might not be able to, and (3) treat that subject matter in an exemplary way. Via the case of the double-slit with single electrons, we have seen how experiments can be considered profound on this view. The double-slit experiment is "about" something profound, it displays its profound content in such a way that brings about a better understanding of quantum theory, and furthermore, it does so because of the way in which it is designed – the experiment captures its profound content in an exemplary way that leads to a positive aesthetic response in those who witness it. Similarly, thought experiments, such as Einstein's elevator, can elicit understanding of some profound content by creating an imaginable scenario that allows us to "see" something that is difficult to grasp in its non-thought experiment form. Further, thought experiments can have aesthetic value: the relation between the content of Einstein's elevator and the way in which it is conveyed through the elements of the narrative brings about an aesthetic response. Thus, we have cases in science which meet the three conditions of profundity proposed by Dodd. In the next section, I turn to a different theory of profundity from philosophy of art to illuminate a different sense in which a scientific experiment can be thought of as "profound".

6.3 Theories of Profundity II: Interpretative Activity

The previous section explored the nature of profound experiments with reference to Kivy and Dodd's analysis of profound artworks. However, a recent paper by Nanay (2021) argues that these accounts of profundity are too permissive, and he proposes a new theory. Here, I do not commit to a particular view of profundity. Rather, I'm interested in exploring the various ways in which experiments can be profound, and Nanay's view helps captures an important source of profundity not yet considered.

Nanay motivates his argument with reference to Damien Hirst's (1991) *The Impossibility of Death in the Mind of Someone Living*. This artwork is comprised of a tiger shark preserved in formaldehyde in a glass case. The title of the work alone indicates that its subject matter is profound and as Nanay notes, it can be said to treat this subject matter in an "exemplary way". That is to say, we can accept that it is an aesthetically valuable, creative, and/or original handling of its profound content. However, Nanay claims that the work is not genuinely profound but rather an exemplar of what he labels the "pseudo-profound". He argues that we need to find a way of separating the former from the latter. Perhaps we might disagree about this particular case, but the point is that we might have artworks that meet the conditions for profundity on Kivy and Dodd's view, yet we would not want to class as profound. Thus, Nanay argues, it is not so apparent that these conditions are necessary for profundity.[5]

In light of this, Nanay offers an alternative proposal for profundity in artworks. Rather than responding to Kivy's work by broadening the "aboutness" of a work beyond what can be expressed in propositions, he abandons the "aboutness" condition to offer a new account that, he argues, is more in tune with our ascriptions of profundity to artworks. Nanay highlights how, both in the history of philosophy and by great writers such as Musil and Proust, profundity has been linked with "darkness" and "unclarity" (2021, 347). Nanay uses these insights as a starting point for his account. Drawing on Nietzsche's example of knowing that a pool of water is deep because we cannot see to the bottom of it, Nanay states 'A literary work or musical composition is profound if it is difficult to see how it works, if it is not clear what is going on, or if the feeling of fluency is missing' (ibid.). On his alternative proposal, a work is profound when it:

1 'actively challenges any straightforward interpretative activity' and
2 'at the same time nudges you to keep trying to interpret it' (ibid.).

While Nanay is also interested in the case of "pure music" which prompted the debates on profundity in philosophy of art, we can see that he is also concerned with the nature of profundity more generally. Hirst's work, for Nanay (who takes it to be an uncontroversial case of the "pseudo-profound"), fails to meet these conditions, especially condition (2).

How about in the context of science and specifically, experimental results? Might we also have experiments that are "pseudo-profound"? One set of experiments that we may class in those terms is those that have results which, if valid, would have been transformative for science, but were ultimately misleading. One example, discussed by Riston in her chapter in this volume, is the detection of a "faster than light" neutrino at CERN which turned out to be the product of a faulty experimental set-up, including a loose fibre optic cable and a clock that ticked too fast. Another candidate is Libet's experiments on free will. Unlike the neutrino example, these were taken to provide new evidence and became widely influential. The experiments appeared to show an astonishing result: there is a delay between the brain "registering" decisions to make certain movements, e.g., pressing a button, and conscious awareness of the decision (Libet 1985; 1999). These experiments have been heavily criticised for their flawed methodology and the assumptions made when inferring from the experimental results to the theoretical conclusions regarding free will.[6] Other candidates for pseudo-profound experiments are those that have surprising results but do not prompt interpretive work. This might be because the results are "merely surprising" to use Morgan's (2005) terminology; while there might be some unexpected results from an experiment, sometimes this can ultimately be explained by existing theory. The more interesting cases of surprise, what Morgan labels "confoundment", cannot be assimilated into existing theory.[7]

To understand how we can have genuinely profound experiments on Nanay's view of profundity, we need to think more about the relationship between experiment and theory.

In his *Representing and Intervening* (1983), Hacking states that philosophers and historians of science have heavily focused on theories, rendering experiments severely overlooked. He sets out to 'initiate a Back-to-Bacon movement, in which we attend more seriously to experimental science' (ibid., 150). Hacking considers experiments in scientific practice, developing the metaphor of the "life" of an experiment to capture how experimentation is more than a practice dictated by theory. This view of experimentation is exemplified in the work of Popper, who states that theoreticians propose well-formed questions to the experimenter whose role is to provide definite answers to those questions and thereby assess the validity of a theoretician's claim. In this view theory 'dominates the experimental world from its initial planning up to the finishing touches in the laboratory' (Popper 1934, 107). Hacking grants a weak claim – experiments should be conducted in light of some understanding or some ideas about the experimental set-up – but sets out to undermine Popper's strong version. For Hacking, "theory" is thus more than mere ideas surrounding an experiment and its apparatus.

There are many dimensions along which an experiment can be said to have a "life". I will focus on those that help us get clearer on the application of Nanay's theory of profundity to scientific experiments. One way in which we can understand this second definition of profundity with reference to the "life" of experiments is by bringing together Nanay's definition of profundity with Morgan's work on surprise and confoundment, mentioned above. Genuinely profound experiments are those in which the results are difficult to understand through existing theory and they "nudge" the scientific community to develop theories in order to try to interpret the results. Unlike our cases of the "pseudo-profound", then, the experiments are neither merely surprising nor are they unreliable or unrepeatable.

An exemplar of this is Ørsted's 1820 discovery of an electromagnetic effect. The set-up of Ørsted's experiment consisted of a wire placed parallel to and over a compass with a magnetic needle. When an electric current was passed through the wire, the needle was deflected from its north-south position. Ørsted conducted variations of this experiment, observing that the direction of the current affects the direction that the needle moves as does the positioning of the wire, i.e., whether it is positioned above or below the needle.

In *Exploratory Experiments: Ampère, Faraday and the Origins of Electrodynamics,* Steinle describes the impact this experiment had:

> 'Ørsted's discovery had been a sensation, sparking far-reaching research efforts across Europe. Almost no one had been prepared for an electromagnetic interaction, and, though none had any trouble replicating the effect, most researchers were left profoundly puzzled ... They simply lacked the concepts and language needed in order to coherently express their results and the complex relationships among them. In particular, the experimental results resisted formulation in terms of the concepts of attraction and repulsion, at least at first blush. But these were the very concepts that had shaped scientific thought at a fundamental level, and not only in physics'
> (2016, 8)[8]

The significance of Ørsted's experiment was widely recognised and produced an entirely new field of research. As Steinle notes, this kind of 'frantic episode' is a rare event in the history of science (ibid., 52). Most notably, the experiment spurred the research of Ampère, a physicist and mathematician based in Paris, and Faraday, a scientist based in London who was, at the time, mainly known for his work in chemistry. Ampère and Faraday both continued Ørsted's experimental work to understand the relation he had uncovered. This led to their development of two competing theories, and then to Maxwell's equations.

While profundity has not received attention in aesthetics of science, there is some work on profound mathematics. Like Nanay, Granger associates profundity with "depth", arguing that one way in which mathematical results are profound or deep is when they elicit a surprising or "odd looking" discovery that 'opens the perspective of creating a new field of objects, which explains this anomaly by "profoundly" renewing the ground upon which the previous theory rest' (1997; 90). While there is more to be said regarding the comparisons between profound experiments and mathematics, this view of what makes for profound mathematics clearly chimes with the features of profound experiments discussed here.

In the case of Ørsted's experiment, we have this surprising new phenomenon, revealing a connection between what were previously considered two separate physical domains. As Steinle explains, there was no clear sense of how to conceptualise and understand this effect, thus it necessitated new theoretical work. This is a case of a profound experiment, but it should be emphasised that Nanay's notion of "interpretive depth" is broader than Morgan's notion of confoundment. We can see this by returning to a previous example, the double-slit experiment with single electrons. This experiment differs from Ørsted's, and it does not fit Morgan's sense of confoundment, but it is not subject to one clear interpretation and "nudges" interpretative work: the experiment is analysed in different ways via the various interpretations of quantum theory. And so, it possesses the key quality of interpretative depth.[9] There are subsequently two ways of understanding Feynman's comment that the experiment contained the "only mystery" of quantum mechanics. The first aligns with our first theory of profundity, i.e., with Dodd's view. The experiment, due to its impeccable design, captures a mysterious phenomenon, and displays its profound content to the viewer in a clear, observable way; a way that would not be possible in normal circumstances. The second aligns with Nanay's view of profundity as discussed in this section: part of the profundity of the experiment is that it presents a mystery; a phenomenon that cannot be straightforwardly interpreted.

What about thought experiments? Can they be profound on this second view of profundity? Hacking (1992) denies that thought experiments can ever have a "life" independent from theory. This would entail that thought experiments also lack this "exploratory" quality that experiments can have. Take Galileo's falling bodies thought experiment in which we are asked to imagine dropping two balls of two different weights from a height, such as the leaning tower of Pisa. What should we expect to see? In the Aristotelian view, the heavier ball should fall faster and the lighter ball should fall slower, as objects fall at a speed proportional to their weight. However, Galileo then asks us to imagine tying the two balls together and dropping this combined object. Again, he asks: what would happen? Galileo shows us that the Aristotelian views provide us with two contradictory answers: (1) the combined object falls faster than the heavier ball on its own (because it's heavier) but also (2) the combined object falls slower than the heavier ball on its own as the lighter ball will pull the heavier back. The thought experiment thus provides reason to reject the Aristotelian view in favour of Galileo's alternative: all objects fall at the same speed.

Although this thought experiment reveals something profound (a contradiction in a long-held theory about the relation between speed and weight) and does so in an exemplary way (it is well designed, even beautifully so) and hence, meets the conditions for profundity on the first view, it has remained "stable", to use Hacking's terminology. Indeed, part of what is aesthetically valuable about this thought experiment is the clarity of its results. Ivanova (2022) discusses "clean" or "solid" results as one way in which experiments can possess beauty, linking this to the notion of a crucial experiment, i.e., one that decides in favour of one theory over another. It appears this source of aesthetic value can be part of our assessments of beauty in thought experiments as well.[10]

However, Hacking is incorrect to claim that all thought experiments lack a "life".[11] Thought experiments can produce results that are puzzling, obscure, and nudge theoretical interpretation, thus producing an aesthetic experience similar to profound artworks. One example is the rocket and thread thought experiment as discussed by Bokulich. This thought experiment was initially devised by Dewan and Beran in 1959 who evaluated the scenario from the perspective of special relativity. It was then used again in 1976 by Bell who did not dispute the result of the thought experiment, but instead showed how the same result could be met by a different theoretical standpoint. Bokulich therefore shows that the relationship between thought experiments and theory is more complicated than Hacking allows for.

Another example, Schrödinger's Cat sticks out as an especially profound or deep thought experiment in Nanay's sense. It was devised in 1935 with the purpose of demonstrating a bizarre feature of the Copenhagen interpretation of quantum mechanics. The set-up of the thought experiment goes as follows: a cat is placed inside a sealed box alongside a small amount of radioactive material and a Geiger counter, connected to a hammer. If radioactive decay is detected, the hammer smashes a bottle of poison, killing the cat. If it's not detected, then the poison is not released and the cat lives. The radioactive material has a 50% chance of decaying over the course of an hour. According to quantum theory, the atom is in a state in which it is both decayed and not decayed. This means the cat is also in a state of superposition: the cat is both alive and dead inside the box. But when the box is opened, an observer will see a definite state: either an alive cat or a dead one. Unlike our central example of a profound ordinary experiment, i.e., Ørsted's discovery of the electromagnetic effect, Schrödinger's Cat did not produce immediate shock or surprise within the scientific community. As French (forthcoming) points out, it wasn't until Feyerabend discussed the example at a conference in the 1950s that it gained significant attention. Since then, the depth of the example has emerged; it has prompted a vast and ongoing discussion in physics and philosophy regarding how to resolve the paradox the thought experiment presents.

Nanay attributes the type of interpretative activity that profound works prompt to features like the length of an artwork, highlighting that we usually attribute "profundity" to long novel. He states how it 'is easier to frustrate straightforward interpretative activities in a 3,000-page novel than in a 3-page short story. This is not to say that it cannot be done in a short story or in a poem, but it is more difficult' (2021, 348). Yet the set-up, for example, of Ørsted's experiment consisted of just a few components and was easy to replicate. Furthermore, Schrödinger's Cat and other scientific thought experiments are short, simple narratives that lack rich detail. Our cases of profound experiments show how profundity in science (and perhaps, then, also in art) can be associated with simplicity.

To sum up so far, while scientists clearly praise parts of scientific practice for their beauty, simplicity, and elegance, we should also think about how a broader set of aesthetic terminology is applied in science. In this chapter, I have focused on the notion of the "profound", an important aesthetic value in the context of art, and its application to experiments. In drawing on two contrasting understandings of profundity, we have seen how theories from philosophy of art may be fruitfully applied to science to further understand our aesthetic appreciation of experiments. Profundity in science has multiple dimensions and can raise interesting questions regarding the relations between our aesthetic evaluation of conventional experiments and other practices in science that are considered "experimental" in some sense.

In his paper on profound artworks, Nanay discusses how his view of profundity challenges a central view of aesthetic experience: the fluency processing account. This view has been invoked in discussions concerning aesthetic values in mathematics and science and is seen as especially insightful in these domains because it offers a way to link the aesthetic with the epistemic. In the final section, I turn to the fluency processing view and its development in the context of aesthetics in mathematics and science. I demonstrate how thinking more about profundity can open interesting areas for future research surrounding "aesthetic-epistemic" feelings in science.

6.4 Conclusion: Profundity, Fluency, and Aesthetic-Epistemic Feelings

According to a central view in psychology developed by, for example, Reber et al. (2004), Reber et al. (2008), Topolinski and Reber (2010), and Schwarz (2018), a distinctive part of aesthetic experience is the "fluency of processing". In this view, aesthetic pleasure is a product of being able to process some object such as a piece of music, a painting, or a mathematical proof, with ease. This is a "metacognitive" feeling in the sense that it is a feeling about our own mental processes. As Reber et al. summarise the view: 'aesthetic pleasure is a function of the perceiver's processing dynamics: the more fluently the perceiver can process an object, the more positive is his or her aesthetic response' (2004, 365). It has been investigated with reference to the way in which both perceptual and conceptual information is processed. The latter is associated with 'the ease of mental operations concerned with stimulus meaning and its relation to semantic knowledge structures' (ibid., 366). Processing fluency increases with features such as symmetry, simplicity, and "goodness of form" as well as with repeated exposure to, or familiarity with, a stimulus.

A particular interesting feature for our purposes is that a cited benefit of the view is that it helps explain the history of the purported connections between beauty on the one hand and truth on the other. This comes up in both the domain of art, captured in statements such as "beauty is truth, truth is beauty" by the likes of Keats, as well as in the domain of science and mathematics. Take, for instance, how Dirac placed 'great confidence in a theory arising from its great beauty, quite independently of its detailed successes' (1980, 40) and Heisenberg's claim that 'if nature leads us to mathematical forms of great simplicity and beauty we cannot help thinking that they are "true"' (1971, 68). Quotations such as these have spurred philosophical interest in aesthetic values in theory selection. Reber et al. cited a range of evidence that evaluations of beauty and evaluations of truth have a common source – processing fluency – which provides an explanation of their association with each other (2004, 377).

For example, Reber et al. (2008) conducted a study into fluency in mathematical reasoning, focusing on whether symmetry is used as a heuristic base for truth in an arithmetic task. Participants were given additions made up of different dot patterns, some of which were symmetric, others asymmetric. Half of the equations provided were correct and the other half were incorrect. Participants had to make a quick decision as to whether the equation was correct, i.e., they were not given enough time to analyse the problem. The study found that there was a higher endorsement of both correct and incorrect equations when the dot pattern was symmetrical. The latter is significant as it is said to rule out the possibility that symmetry aided the calculation (ibid., 12).

The research on fluency has been picked up in philosophical discussions of the aesthetics of mathematics and science by Todd (2017; 2020) and Dutilh Novaes (2019) who are interested in the connections between the aesthetic and the epistemic. In particular, they seek to respond to a sceptical line of thought that apparent aesthetic judgements in the domain of science and mathematics are not really aesthetic ones. Rather, aesthetic language is used in a mere metaphorical way.[12] Dutilh Novaes and Todd each use the fluency account as a basis for their own view, arguing that it helps establish the ways in which epistemic feelings can have an aesthetic quality and vice versa, as well as the possibility of hybrid aesthetic-epistemic feelings (Todd 2017, 227).

Dutilh Novaes focuses on "aha experiences", characterised in terms of fluency. These experiences involve: 'suddenness (the experience is surprising and immediate), ease (the solution is processed without difficulty), positive affect (insights are gratifying), and the feeling of being right (after an insight, problem solvers judge the solution as being true and have confidence in this judgment)' (Topolinski and Reber 2010, 402). Dutilh Novaes argues that surprising mathematical proofs elicit positive affective responses that, in turn, increase their persuasiveness (their feeling of "being right") (2019, 74).

Todd is also interested in "aha experiences" but departs somewhat from the fluency processing view. He argues that the focus on fluency is too narrow, highlighting that there are other aesthetic-epistemic feelings that ought not to be subsumed under fluency, though they might be closely related and have a similar phenomenology. He proposes a broader category, the "feeling of fittingness". While empirical work is yet to be done on such a feeling, Todd proposes that such research should be conducted with the following hypothesis:

> "All epistemic feelings and aesthetic experiences are determinate variations of the 'feeling of fittingness', which is a determinable experiential manifestation of certain sub-personal processes of representing and assessing the relation between certain cognitive processes and certain properties of the stimuli at which they are directed. When this relation exhibits, among other things: a) coherence b) breadth of scope, c) simplicity, and d) fluency, we have what I will call 'cognitive consonance'. It is this consonance that is manifested as a feeling of fittingness, and this feeling is jointly aesthetic-epistemic in nature"
>
> (2017, 229)

In the previous sections, I presented two theories of profundity from philosophy of art that provide a starting point for illuminating the features of profound scientific experiments. We have also seen that profundity in experiments is not just an aesthetic feature – it also has an epistemic aspect in that it either is (a) a way of presenting profound subject matter so that we better understand it, or (b) resists a straightforward interpretation of experimental results while nudging us to find one. Hence, as with the cases that Todd and Dutilh Novaes are interested in, profundity has both an aesthetic and an epistemic component. It is therefore helpful to consider profundity alongside such theories.

The first theory of profundity that I discussed developed by Dodd could be accommodated within these accounts, especially, perhaps, if we accept Todd's broader category of aesthetic-epistemic feelings. In particular, it chimes with Todd's discussion of the feeling of understanding. With reference to the work of Elgin, he discusses how through models, thought experiments, and idealisations in science, we can gain "direct epistemic access" to a phenomenon in a way that makes it easier to understand. He adds that an important aspect of this understanding is affective in nature: it 'frequently has an aesthetic character, a feeling of fit' (2020, 80). The second theory of profundity, however, puts pressure on the scope of these views. As we saw, Nanay's view identifies profundity with a certain effect that artworks can have on us. Profound artworks do not allow for a straightforward interpretation. For this reason, Nanay states, experiencing profundity 'is the opposite of an experience of fluency … profound works actively work against the fluency of processing' (2021, 352). We can add to this that profundity in the second sense also goes against a "feeling of fittingness".

Broadening our accounts of aesthetics in science beyond properties like simplicity to think about "profound" experiments therefore provides an opportunity to develop discussions of aesthetic-epistemic feelings in science. One potentially fruitful direction is to draw from the rich literature on aesthetic emotions like "interest", "wonder", and "awe" and consider them in the context of science (Fingerhut and Prinz 2020). Like profundity in the second sense, these are not (at least not easily) assimilated with feelings of fluency but nonetheless have an epistemic and/or cognitive quality. Another is to discuss sociological studies pointing to the role of various epistemic emotions in science (Kozlov forthcoming) alongside Dutilh Novaes and Todd's defences of aesthetic-epistemic feelings in science and mathematics. As with other qualities flagged up in this volume, such as Parsons on the sublime, Ritson on the value of ugliness, and Ivanova on the plurality of experimental aims, looking at profundity demonstrates that accounts of aesthetics in science need to be able to accommodate cases in which experimental results are obscure, perplexing, and cannot be straightforwardly understood. The tensions between feelings of fluency (and by extension, feelings of fittingness) and the experience of profundity in the second sense outlined thus warrants further investigation.[13] This is especially pertinent when such features encompass a scientific practice's aesthetic and epistemic value as in the case of profound experiments.

Acknowledgements

This chapter was presented at The Aesthetic Value of Scientific Experiments workshop in Cambridge, the BSPS Annual Conference, and EPSA21. I am very grateful to the audience and fellow speakers for their helpful comments and questions. Thank you also to Milena Ivanova, Steven French, and Nathan Wood for their feedback, and to Erik Curiel for a discussion on profundity in science.

Notes

1 Aumann (2014, 117) also includes being true or misleading as potential aesthetic properties that have a cognitive nature. This list of types of aesthetic properties not meant to be complete, and these categories may overlap in various ways. My aim is to show that aesthetic terminology is far broader than what is typically discussed in the literature on aesthetics in science.
2 The discussion of profundity in philosophy of art has been centred around a question that was prompted by Kivy (1990; 1997) of whether "pure music" or "music alone" can be considered profound. For Kivy, while we label certain musical works, such as Beethoven's late quartets as profound, we are mistaken. This is because pure music fails to meet the semantic condition, i.e., condition (1) as it is not *about* anything profound. In fact, it lacks any content at all. In Kivy's view, as Dodd (2014) and Davies (2002) discuss, this "aboutness" condition is a strong one in the sense that it is a matter of having some propositional content. In response, they have expanded what is meant by content beyond Kivy's narrow conception in order to show the ways in which pure music can have content, and therefore can also have profound content and meet the above conditions. Seeing as I am not interested in the case of pure music here, but rather, theories of profundity more generally, I will leave these debates to one side.
3 For experiments conducted in a laboratory, there is work to be done to demonstrate that the system informs us of the world outside the experimental setup. In the process of conducting experiments and analysing their results, scientists need to show why their experimental system is informative of their ultimate target, that is, they need to establish its external validity. This is the subject of extensive discussion in the epistemology of experiment and science studies more generally. It is not a simple case of a scientist having immediate access to the target. See Cartwright (2007), Currie and Levy (2019), Morgan (2005), Parker (2009), and Roush (2018) for discussion.
4 For an account of the relation between form and content as an aesthetic property which has epistemic value in science, see Murphy (forthcoming). This is a broader view than discussed here, i.e., it is not concerned with profundity.
5 Nanay takes Kivy and Dodd's views together when making this argument. However, it could be argued that Hirst's work fails to be profound on Dodd's view because it does not handle its subject matter in such a way that brings

about a deeper understanding of it. As I am interested in the different perspectives on profundity rather than the details of this specific case, I will leave this to one side.

6 Despite criticism, Libet's experiment is taken to raise a number of interesting questions. For a philosophical discussion of this experiment, see Bayne (2011).

7 For more on the value of surprise in science, see Parke (2014), Currie (2018), Ritson (2020), and French and Murphy (2021).

8 For Steinle, this counts as an "exploratory" experiment, i.e., one that brought about "conceptual insecurity", "even speechlessness" by having unexpected results 'that, while clearly observable and reliably repeatable, remain resistant to treatment by means of customary concepts and are sometimes frankly ineffable' (2016, 6).

9 Another sense in which Nanay's view of profundity when applied to scientific experiments comes apart from Morgan's notion of confoundment will become apparent in the final section: profundity is an experiential concept and can thereby be discussed alongside aesthetic and epistemic emotions. See Currie (2018) for a discussion of Morgan's view of surprise as an "epistemic good" which he contrasts with a psychological notion.

10 For a discussion of thought experiments and crucial experiments, see Bokulich (2001).

11 Here I focus on the "life" of thought experiments in terms of their interpretive depth. For more on the various ways thought experiments can have a "life" in Hacking's sense of the term, including how they can precede theory, see Shinod (2017) and Murphy (2020b). For a discussion of how experiments can be reconsidered in light of different theories than they were initially taken to test see Patton (2011).

12 For other detailed discussion of this line of thought, see Todd (2008), O'Loughlin and McCallum (2019), and Murphy (forthcoming).

13 See also Arcangeli and Dokic for a discussion of the sublime in science and its tension with the fluency processing accounts (2020, 117).

Bibliography

Arabatzis, T. (2008) 'Experiment', in S. Psillos and M. Curd (eds.) *The Routledge Companion to Philosophy of Science*, 1st edition. London: Routledge, pp. 159–170.

Arcangeli, M. and Dokic, J. (2020) 'A Plea for the Sublime in Science', in M. Ivanova and S. French (eds.) *Aesthetics of Science: Beauty, Imagination and Understanding*. New York: Routledge.

Aumann, A. (2014) 'The Relationship between Aesthetic Value and Cognitive Value', *The Journal of Aesthetics and Art Criticism*, 72(2), 117–127.

Bayne, T. (2011) 'Libet and the Case for Free Will Scepticism', in R. Swinburne (ed.) *Free Will and Modern Science*. Oxford: OUP/The British Academy. pp. 25–46.

Bird, A. (2020) 'How Can Loveliness Be a Guide to Truth? Inference to the Best Explanation and Exemplars', in Milena Ivanova and Steven French (eds.) *Aesthetics of Science: Beauty, Imagination and Understanding*. New York: Routledge. https://www.taylorfrancis.com/chapters/edit/10.4324/9780429030284-7/loveliness-guide-truth-inference-best-explanation-exemplars-alexander-bird

Bokulich, A. (2001) 'Rethinking Thought Experiments', *Perspectives on Science*, 9(3), 285–307.

Carroll, N. (1999) *Philosophy of Art: A Contemporary Introduction.* London: Routledge.

Cartwright, N. (2007) 'Are RCTs the Gold Standard?', *BioSocieties*, 2, 11–20.

Crease, R. (2002) 'The Most Beautiful Experiment', *Physics World*, 15(9), 19–20.

Currie, A. (2018) 'The Argument from Surprise', *Canadian Journal of Philosophy*, 48(5), 639–661.

Currie, A. (forthcoming) 'Epistemic Engagement, Aesthetic Value & Scientific Practice', *British Journal for the Philosophy of Science.*

Currie, A. and Levy, A. (2019) 'Why Experiments Matter', *Inquiry*, 62(9–10), 1066–1090.

Davies, S. (2002) 'Profundity in Instrumental Music', *British Journal of Aesthetics*, 42, 343–356.

Dirac, P. (1980) 'The Excellence of Einstein's Theory of Gravitation', in M. Goldsmith, A. Mackay and J. Woudhuysen (eds.) *Einstein: The First Hundred Years.* Oxford: Pergamon Press. pp. 11–14.

Dodd, J. (2014) 'The Possibility of Profound Music', *The British Journal of Aesthetics*, 54(3), 299–322.

Dutilh Novaes, C. (2019) 'The Beauty (?) of Mathematical Proofs', in A. Aberdein and M. Inglis (eds.) *Advances in Experimental Philosophy of Logic and Mathematics.* London: Bloomsbury Academic, pp. 63–93.

Edge.org: (2012) 'What is your favorite deep, elegant or beautiful explanation?' https://www.edge.org/annual-question/what-is-your-favorite-deep-elegant-or-beautiful-explanation. Accessed Oct 23, 2022.

Elgin, C. Z. (1993) 'Understanding: Art and Science', *Synthese*, 95(1), 13–28.

Feynman, R. (1965) Lectures on Physics.

Fingerhut, J. and Prinz, J. (2020) 'Aesthetic Emotions Reconsidered', *The Monist*, 103(2), 223–239.

French, S. and Murphy, A. (2021) 'The Value of Surprise in Science', *Erkenntnis.* https://doi.org/10.1007/s10670-021-00410-z

Goodman, N. (1976) *Languages of Art*, 2nd edition. Indianapolis, IN: Hackett.

Goodman, N. (1985) 'How Buildings Mean', *Critical Inquiry*, 11, 642–653.

Granger, G. G. (1997) 'What Is a Profound Result in Mathematics?', in E. Agazzi and G. Darvas (eds.) *Philosophy of Mathematics Today. Episteme*, vol 22. Dordrecht: Springer. pp. 89–100.

Hacking, I. (1983) *Representing and Intervening: Introductory Topics in the Philosophy of Natural Science.* Cambridge: Cambridge University Press.

Hacking, I. (1992) 'Do Thought Experiments Have a Life of Their Own? Comments on James Brown, Nancy Nersessian and David Gooding'. *PSA: Proceedings of the Biennial Meeting of the Philosophy of Science Association*, pp. 302–308.

Heisenberg, W. (1971) *Physics and Beyond.* London: Allen & Unwin.

Ivanova, M. (2021) 'The Aesthetics of Scientific Experiments', *Philosophy Compass*, 16(3).

Ivanova, M. (2022) 'What Is a Beautiful Experiment?', *Erkenntnis.* https://doi.org/10.1007/s10670-021-00509-3

Kivy, P. (1990) *Music Alone: Philosophical Reflections on the Purely Musical Experience.* Ithaca and London: Cornell University Press.

Kivy, P. (1997) 'The Quest for Musical Profundity', in *Philosophies of Arts: An Essay in Differences.* Cambridge: Cambridge University Press, pp. 140–178.

Kozlov, A. (forthcoming) 'Scientific Experiments Beyond Surprise and Beauty'.

Libet, B. (1985) 'Unconscious Cerebral Initiative and the Role of Conscious Will Involuntary Action', *Behavioral and Brain Sciences*, 8, 529–566.

Libet, B. (1999) 'Do We Have Free Will?', *Journal of Consciousness Studies*, 6(8–9), 47–57.

Morgan, M. S. (2005) 'Experiments versus Models: New Phenomena, Inference and Surprise', *Journal of Economic Methodology*, 12(2), 317–329.

Murphy, A. (2020a) 'The Aesthetic and Literary Qualities of Scientific Thought Experiments', in M. Ivanova and S. French (eds.) *Aesthetics of Science: Beauty, Imagination and Understanding.* New York: Routledge. pp. 146–167.

Murphy, A. (2020b) 'Thought *Experiments and the Scientific Imagination'.* PhD Thesis, University of Leeds. https://etheses.whiterose.ac.uk/27984/

Murphy, A. (forthcoming) 'Form and Content: A Defence of Aesthetic Values in Science'. *Philosophy of Science.* https://doi.org/10.1017/psa.2023.46

Nanay, B. (2021) 'Looking for Profundity (in All the Wrong Places)', *The Journal of Aesthetics and Art Criticism*, 79(3), 344–353.

Nguyen, C. T. (2020) *Games: Agency as Art.* New York: Oxford University Press.

O'Loughlin, I. and McCallum, K. (2019) 'The Aesthetics of Theory Selection and the Logics of Art', *Philosophy of Science*, 86(2), 325–343.

Parke, E. (2014) 'Experiments, Simulations, and Epistemic Privilege', *Philosophy of Science*, 81(4), 516–536.

Parker, W. (2009) 'Does Matter Really Matter: Computer Simulations, Experiments, and Materiality', *Synthese*, 169, 483–496.

Parsons, G. G. and Rueger, A. (2000) 'The Epistemic Significance of Appreciating Experiments Aesthetically', *British Journal of Aesthetics*, 40(4), 407–423.

Patton, L. (2011) 'Reconsidering Experiments', *Hopos: The Journal of the International Society for the History of Philosophy of Science*, 1(2), 209–226.

Popper, K. (1934/2002) *The Logic of Scientific Discovery.* London and New York: Routledge.

Reber, R., Brun, M. and Mitterndorfer, K. (2008) 'The Use of Heuristics in Intuitive Mathematical Judgment', *Psychonomic Bulletin & Review*, 15(6), 1174–1178.

Reber, R., Schwarz, N. and Winkielman, P. (2004) 'Processing Fluency and Aesthetic Pleasure: Is Beauty in the Perceiver's Processing Experience?', *Personality and Social Psychology Review*, 8(4), 364–382.

Ritson, S. (2020) 'Probing Novelty at the LHC: Heuristic Appraisal of Disruptive Experimentation', *Studies in History and Philosophy of Science Part B: Studies in History and Philosophy of Modern Physics*, 69, 1–11.

Roush, S. (2018) 'The Epistemic Superiority of Experiment to Simulation', *Synthese*, 195(11), 4883–4906.

Schrödinger, E. (1955) 'The Philosophy of Experiment', *Nuovo Cim*, 1, 5–15.

Schwarz, N. (2018) 'Of fluency, beauty, and truth: Inferences from metacognitive experiences', in J. Proust & M. Fortier (eds.), *Metacognitive diversity: An interdisciplinary approach.* Oxford: Oxford University Press, pp. 25–46.

Shinod, N. K. (2017) 'Why Thought Experiments Do Have a Life of Their Own: Defending the Autonomy of Thought Experimentation Method', *Journal of Indian Council for Philosophical Research*, 34(1), 75–98.

Steinle, F. (2016) *Exploratory Experiments: Ampère, Faraday, and the Origins of Electrodynamics*. Pittsburg, PA: University of Pittsburg Press.

Todd, C. (2008) 'Unmasking the Truth Beneath the Beauty: Why the Supposed Aesthetic Judgements Made in Science May Not Be Aesthetic at All', *International Studies in the Philosophy of Science*, 22(1), 61–79.

Todd, C. (2017) 'Fitting Feelings and Elegant Proofs: On the Psychology of Aesthetic Evaluation in Mathematics', *Philosophia Mathematica*, 26(2), 211–233.

Todd, C. (2020) 'Imagination, Aesthetic Feelings, and Scientific Reasoning', in M. Ivanova and S. French (eds.) *The Aesthetics of Science; Beauty, Imagination and Understanding*. New York: Routledge.

Topolinski and Reber (2010) 'Gaining Insight Into the "Aha" Experience', *Current Directions in Psychological Science*, 19(6), 402–405.

Turner, D. (2019) *Paleoaesthetics and the Practice of Paleontology*. Cambridge: Cambridge University Press.

Wragge-Morley, A. (2020) *Aesthetic Science: Representing Nature in the Royal Society of London, 1650–1720*. Chicago and London: The University of Chicago Press.

7 Experimental Science as Epistemic Expansion

New Work for a Theory of the Sublime

Glenn Parsons

Philosophical discussions of aesthetics and science have tended to focus on the relationship between certain "aesthetic qualities" – notably beauty, simplicity, and elegance – and the traditional epistemic aims of science, such as truth, empirical adequacy, explanation, understanding, and so on.[1] Many commentators have wondered whether, and how, preferring theories with these qualities might advance said epistemic aims. Could beauty lead us towards truth, for example? This focus on so-called "classical" aesthetic qualities and epistemic aims holds not only in discussion of scientific *theories* (which has dominated discussions of aesthetics and science) but also in discussion of the aesthetics of *experiments* as well.[2] In this chapter, I take a different angle on the latter topic, exploring a less canvassed and decidedly "non-classical" aesthetic notion: the sublime.[3]

Traditionally, the idea of the sublime found its main application in 17th and 18th-century aesthetics of nature. Standard sublime fare included the starry heavens, rushing waterfalls, lofty mountain peaks, and other overwhelming scenes. I argue here, however, that the sublime can today find application in accounting for certain aesthetic aspects of experimental science that are important but underdescribed, partly for want of an appropriate aesthetic vocabulary with which to describe them. I claim that the sublime, suitably reinterpreted for our times, provides our best way of explicating this important aesthetic dimension of science. In making this case, I will take issue with other recent revivals of the sublime that take the concept in a different, and somewhat more traditional, direction.

The plan of the chapter is as follows. I begin with a brief sketch of the sublime's eventful history. Once a significant, indeed central, concept in European aesthetics, it subsequently fell into neglect in philosophy but has recently seen renewed interest. In Section 7.2, I outline the standard view of sublime experience that has emerged in these new accounts, one centred on the notion of *cognitive failure*. In Section 7.3, I turn to experimental science and, in particular, to the use of certain observational techniques, describing what I take to be an important aesthetic dimension of their use. This dimension is most aptly captured,

DOI: 10.4324/9781003263920-8

I argue, in terms of the sublime. However, since the standard account fails to apply to these cases, in Section 7.4, I propose an alternative analysis of the sublime in terms of *epistemic expansion*. This account not only fits the sublime's history, accounting for canonical examples of sublime experience but also applies to the relatively new and neglected kinds of experimental experiences that I have described. I conclude, in Section 7.5, with some implications of this proposed redeployment of the sublime for two more general issues: the role of philosophy in aesthetic discourse, and the scope of the aesthetics of experiment as a domain of study in philosophical aesthetics.

7.1 The Rise and Fall of the Sublime

Historians of Western culture tell us that until the mid-17th century, there is little sign of positive appreciation of disorderly and overwhelming landscapes: mountains, deserts, tangled forests, swamps, and so on. In the ancient and medieval periods, appreciation seems to cleave to the cultivated, humanised world of orchard, pasture, and farm.[4] By the 17th century, however, Europeans began to seek out and appreciate, rather than avoid and scorn, wilder places. This shift in attitude is evident in the rise of landscape painting as an artistic genre: by the mid-17th century, painters like Salvator Rosa were starting to incorporate wilder landscapes in traditional historical painting. A century or so later, foreboding, frightening landscapes, such as alpine scenery, had become an artistic subject in their own right.[5]

With this shift in landscape Tastes, however, Europeans' aesthetic experiences outstripped their aesthetic vocabulary, since the traditional label "beautiful" carried connotations of harmony, order, proportion, and pleasure from its classical origins.[6] It, therefore, was an awkward fit for these new sites of a darker and more difficult aesthetic experience. As a way of describing these new experiences, 17th-century writers turned instead to another term with roots in antiquity – the sublime.

Most of us know the term "sublime" from a mundane use in contemporary English, where it simply means "elevated or great", as in the adage "from the sublime to the ridiculous", or commonplace uses like "That custard was sublime". But this humble role belies a grand and eventful history. The term originates in ancient rhetoric, being the subject of the first century CE treatise known as the *Peri Hupsos*, in which it is applied primarily to certain forms of speech or writing. Though there were prior translations, the text began to garner attention in European criticism after Nicholas Boileau translated it into French in 1674. Seventeenth-century writers shifted the term's use somewhat, employing it to designate not only speech or writing, but also elevated or great *natural objects*, such as mountains, the sea, storms, and the like, as well as the elevated *state of mind, or feeling,* that they produced.[7]

The introduction of the term was spectacularly successful, and the sublime not only took its place alongside beauty in European literary discourse but in fact elbowed the beautiful aside. It chimed well with the Romantic movement's belief in the special significance of the wilder, transcendent side of nature. A characteristic example is Wordsworth's "Tintern Abbey", in which he wrote of

> a sense sublime
> Of something far more deeply interfused,
> Whose dwelling is the light of setting suns,
> And the round ocean and the living air,
> And the blue sky, and in the mind of man:
> A motion and a spirit, that impels
> All thinking things, all objects of all thought,
> And rolls through all things.[8]
>
> Wordsworth (1798)

But even as poets were waxing Metaphysical, the sublime was also attracting more sober attention from philosophers.[9] Burke and Kant, most notably, explained sublime experience, not in terms of contact with a transcendent spirit, but in terms of a specific psychological response. For Burke, vast and powerful objects, such as an alpine avalanche, inspire a terror that arrests the mind. But when observed from a safe distance, where we need not fear for our personal safety, this terror is transformed into what Burke calls 'a sort of delightful horror … a tranquility tinged with terror'. It is this 'mixed feeling' that we call sublime, according to Burke (1757, Pt. IV, sec. VII).

A few decades after Burke, Kant offered a more complicated account on which vast or powerful objects make us feel overwhelmed and frustrated, thereby inspiring some form of positive reflection on the self. When we look at the Milky Way overhead, for example, we cannot imagine or perceive it all – it is simply too big for our faculties to grasp. This negative feeling of frustration at our limitations is offset, however, by the realisation that we do, after all, possess a *rational* faculty that *can* grasp the scale of the heavens mathematically. This recognition that this rational faculty is absent in nature, and hence superior to it, is the source of our pleasure in the sublime, for Kant. (This leads Kant to say, notoriously, that what is truly sublime is not external nature, but we ourselves) (1790).[10]

There are certainly substantial differences between the theories of Burke and Kant, and later philosophers offered still other accounts. However, it is worth noting that, from a more coarse-grained perspective, these accounts are quite similar in outline: for each, sublime experience does not involve contacting mystical spirits, as Wordsworth's treatment suggested. Rather, for both Burke and Kant, sublime experience principally involves a certain kind of *cognitive failure*: an arrest of the rational faculties of the mind that, due to special circumstances, ultimately results in a distinctive sort of pleasure.

Like the other major conceptual innovation of 18th-century European aesthetics, the Picturesque, the sublime's star eventually faded in popular tastes. And after Kant, philosophers tended to assign it a diminished role.[11] Nonetheless, the concept did not die; in fact, it subsequently found wide application in other areas, including theorising about the experience of Art. Indeed, theories of the sublime are now legion: writers have discussed the Technological sublime, the Political sublime, the Feminine sublime, and many more besides. The notion's malleability has long been a source of philosophical worry: writing in 1904, E. F. Carritt already worried that theorising had left the sublime 'so vague, and the varieties of it so heterogenous that we may ask whether it is real class at all' (1904, 357).[12] Echoing Carrift's frustrations, the literary critic James Elkins recently proposed a moratorium on the term, as one of those overused words that 'have been said without thought so many times that they are like blank coins, rubbed by thousands of fingers until they are nothing but thin blank disks' (2011; reprinted in Clewis 2019a).

7.2 Recent Reconsiderations: The Standard View

By the early 20th century, the sublime had largely fallen off the radar of Anglo-American philosophy. In the early 1990s, Guy Sircello observed that 'there is virtually no philosophical literature on this topic ... since the classic days' of Burke and Kant. Around the same time, Ronald Hepburn put the question bluntly in an essay entitled 'The Concept of the Sublime – Has it any Relevance for Philosophy Today?' (Hepburn 1988; Sircello 1993, 542).

Both Hepburn and Sircello answered this question affirmatively, arguing that, despite the "overly speculative" flavour of much theorising about the sublime, the concept could still play an important role in describing certain aesthetic experiences.[13] Analysing these descriptions, Sircello identified, as central to sublime discourse, an idea that he called the "theme of epistemological transcendence". This is the idea that 'human mental powers in general are revealed [in experience of the sublime] to have radically limited access to ... "reality"' (1993, 543). Sircello admits that some ways of explicating this theme land us in incoherence. For instance, talk of sublime experience bringing us into contact with an "Unknowable Reality" is simply contradictory, since this involves knowing the unknowable.[14] But he also suggests that 'the theme of epistemological transcendence may simply mean that for any possible given set of routes to epistemological access to "reality," that set is insufficient to provide a complete understanding or grasp of "the real"' (ibid.). In other words, sublime experience involves becoming aware that an object cannot be *completely* known. Another way to put the point, though not Sircello's, is that the sublime involves an encounter with an object that triggers awareness of a profound kind of *cognitive failure or limitation*.[15]

Sircello's analysis is reminiscent of the traditional accounts of Kant and Burke, which both involve an arrest or stoppage of the mind (Kant's notion of the mathematical sublime, in particular, appeals to the recognition of just this sort of cognitive limitation). More recent attempts by philosophers to revive the sublime have followed him in returning to the quite traditional notion of cognitive failure. Emily Brady's (2013) book, *The Sublime in Modern Philosophy*, is a sustained defence of a broadly Kantian approach involving 'an intense emotional response distinguished by feelings of being overwhelmed – a somewhat anxious experience, but ultimately one that feels exciting and pleasurable' (2013, 117–118).[16] More recently, Aracangeli and Dokic have offered a sophisticated account of sublime experience that builds on recent work in the psychology of awe and related emotions, but which also cleaves to a broadly Kantian approach. A closer look at their detailed account will furnish a useful illustration of the contemporary approach.

In their seminal paper on the psychology of awe, the psychologists Keltner and Haidt characterise emotion as involving two elements. One is 'perceived vastness', which need not be spatial, but can be 'anything that is experienced as being much larger than the self' (Keltner and Haidt 2003, 303). The second is a felt need for accommodation, which is a 'process of adjusting mental structures that cannot assimilate a new experience'. This need for accommodation is prompted by 'a challenge to or negation of mental structures when they fail to make sense of something vast'. This challenge, they hold, produces a negative feeling, as 'such experiences can be disorienting and even frightening ... since they make the self feel small, powerless, and confused' (ibid.).

Building on this general account of awe, Arcangeli and Dokic (2021) propose that the sublime represents one particular form of awe.[17] They characterise sublime experiences as involving both a negative and positive element. First, an encounter with an overwhelmingly vast or powerful object triggers what they call a "radical limit experience", in which the perceiver becomes aware that they have reached a global or absolute cognitive limit. They explain this idea by way of a contrast with a mundane-limit experience, such as hearing the highest note that the human ear can register. In such an experience, one becomes aware that one is operating at the limit of hearing, but this limit is not a global or absolute limit, since the subject can imagine or at least conceive of sounds beyond this limit. In a *radical* limit experience, in contrast, the subject becomes aware that there is no other way in which their cognitive powers can accommodate the object: the felt cognitive failure is therefore "global". Arcangeli and Dokic illustrate with the traditional example, employed by Kant, of viewing the starry heavens: when the visible universe is too big to perceive, and the agent becomes "aware that she cannot do better by any other cognitive means", a radical limit experience is triggered.

This failure produces, first, a "feeling of self-negation: the subject does not feel part of the world anymore". Because the subject's faculties are unable to grasp the object, the subject has the feeling that they are "wholly disappearing from the world". This feeling of dislocation or disappearance of the self represents the negative side of the sublime experience (2021, 154–155).[18]

However, in line with the general structure of awe as outlined by Keltner and Haidt, this perceived global cognitive failure also gives rise to a desire for accommodation. In sublime experience, this is achieved when 'the subject realises that there is a cognitive refuge after all. She succeeds in mobilising novel or unsuspected cognitive resources that are not concerned by the limit-experience' (2021, 158). So in the example of viewing the starry heavens, the subject overcomes the initial feeling of cognitive failure by 'by bringing in a sophisticated concept of infinity, which enables her to make sense of what she is confronted with' (ibid.). This allows the subject to feel part of the world once more and represents the positive, pleasurable, aspect of sublime experience.

Like other contemporary views of the sublime, Arcangeli and Dokic's differs in many points of detail from the classical theories of Kant and Burke. Nonetheless, all these views share the same basic conception of sublime experience: a vast or overwhelming natural object triggers a temporary and distressing cognitive failure that is then overcome through some kind of self-reflection. This approach is sufficiently common among the sublime revivalists that it may justly be called the "standard philosophical account" of the sublime.

7.3 Inherently Prosthetic Perception and the Aesthetics of Experiment

In this section, I shift focus to examine a kind of experience that has played an important role in experimental science, and that I think should move us to *resist* the standard account of the sublime sketched in the previous section. The kind of experience I have in mind is hardly esoteric: it is an experience in which various technologies and techniques have been used to expand the range of phenomena that we can perceive in the natural world.

In the history of scientific observation and experiment, techniques for perceptual expansion have come in many different forms, the most obvious being the use of simple optical instruments, such as telescopes. Only after Galileo trained his telescope on Jupiter were we able to perceive its moons, overcoming the light-gathering limitations of the human eye. Other technologies have transcended more fundamental limitations. Light microscopy, for instance, allowed us to resolve objects too small to be seen with the unaided eye. The early images produced by Hooke and other pioneers of microscopy therefore represented the unlocking of an entirely new realm of experience.

Beyond the telescope and microscope, there are many other examples of experimental techniques for expanding perception, such as:

- Cloud chambers, once widely used in physics, allow for the observation of the movement of subatomic particles (a chamber is filled with a supersaturated vapour, and as energetic particles pass through it, they leave visible condensation trails in the vapour).
- Immunofluorescence labelling, employed in cell biology, allows for the visualisation of specific proteins at the cellular level (specially engineered antibodies are attached to specific proteins in the cell and emit coloured light when irradiated, allowing experimenters to see where a particular kind of molecule is localised with an individual cell).
- Scanning Electron Microscopy, which allows us direct visualisation of entities that are too small to resolve with light (a sample of material is bombarded with a beam of electrons: scattered electrons are then used to generate a three-dimensional image of the sample structure).
- Digital astrophotography, which allows for the visualisation of deep space objects, such as nebulae, huge clouds of interstellar gas and dust, that are too dim to show detail even in a powerful optical telescope (through digital astrophotography, astronomers can produce extremely detailed photographs of the fine structure of these objects by collecting, in long exposures, far more light than the human eye is capable of collecting).[19]

A more recent example can be found in the creation of the so-called "first direct image" of a black hole in 2019 by the Event Horizon Telescope, a network of radio wave telescopes spread across the globe. Though black holes cannot be seen, since they emit no light, the telescope array was able to produce an image of the accretion disk, a swirling disk of heated matter surrounding the event horizon, of the black hole at the centre of galaxy M87.[20]

Despite the wide array of different techniques for achieving perceptual expansion, it will be useful to have a blanket term that applies to all of them. So I will call the new kinds of perception that these techniques made possible *inherently prosthetic perception*, or "IPP" for short. The idea is that these are perceptual experiences that, given the limitations of the human perceptual system, are not possible without the use of the relevant technologies (telescopes, microscopes, and so on). The black hole in M87, the fine structure of the Eta Carina Nebula or the Drosophila eye, the movement of helium nuclei, the location of the tubulin protein in a given cell: none of these can be perceived by humans, given the kind of perceptual systems we possess. Yet each of these is something that we now can not only *know* about but also *perceive* – actually see – via the imaging techniques I described.[21]

As mentioned, this is certainly not an unfamiliar phenomenon. However, it has not been given much attention from the *aesthetic* point of view, and regarding this, I want to make two claims.

First, I want to suggest that the experience of IPP is an aspect, though a hitherto neglected one, of the aesthetics of experimental science. Thus, I claim that, in addition to its obvious epistemic role, this kind of experience also has an aesthetic dimension. Now, the sorts of experiments and observations that I have described might involve many different kinds of aesthetic value. For instance, one might admire the image of the M87 black hole simply as an abstract pattern, for its colours and shapes. But the aesthetic value that I have in mind is one that arises from the fact that, in this image, we are *seeing* an object that, in the normal course of things, cannot be seen. Another way to put the point is that the aesthetic value I have in mind is one that involves the particular sort of *representational content* possessed by the experience in question.

How contentious is the claim that the experience of IPP can involve this sort of aesthetic value? That will depend on how we understand the aesthetic, a philosophically controversial matter. I favour an inclusive, and still fairly prominent, account, which holds that something has aesthetic value when experiencing it is valued for its own sake, and not for the sake of some further end or benefit that the experience brings.[22] Since this view does not link the aesthetic to any specific features of objects or experiences, it allows that more or less any kind of thing can, in principle, have aesthetic value, so long as experience of it is valued for its own sake. Techniques for facilitating IPP obviously have value for the knowledge they provide, and this epistemic value is our chief reason for inventing and pursuing them. However, if the experience that they provide – IPP – is also valued just for its own sake – simply as a rewarding kind of experience to have – then they will have aesthetic value as well.[23]

But is it so valued? Does the scientist, or anyone else for that matter, find the experience of IPP to be a kind of experience that is rewarding for its own sake? While one can point to scattered remarks pointing in this direction (see below for some examples), I have no rigorous psychological data to support such an empirical claim. Furthermore, the average scientist is reticent about such matters (a point to which I shall return below) and from what they do say, it is, frankly, hard to draw any firm conclusions about the precise reasons they have for valuing this or that experience. But whether IPP is, as a matter of act, valued aesthetically or not, I do claim that it is at least aesthetically *valuable* (which, as noted, is quite a weak claim, at least on my preferred account of the aesthetic). And here I will fall back on my own experience in asserting that there is still, three centuries on from Hooke, something visually captivating and compelling, on a human level, about peering into the hidden crevices of nature in this way.[24]

7.4 Reconceiving the Sublime as Epistemic Expansion

If it is granted that our experiences of IPP can have the aesthetic dimension described, I want to advance a second, and more contentious, claim: this aesthetic dimension is best understood – most aptly described – as an experience of the *sublime*.

This is a more contentious claim because IPP, as I have described it, does *not* fit with the standard account of the sublime described in Section 7.2. Central to that account is a feeling of profound cognitive failure or frustration resulting from an inability to comprehend or perceive what we are encountering. But when I see a digital astrophotograph of the Carina nebula, for instance, I am not frustrated at not being able to see the nebula, since I *do* see it. Nor, typically, does IPP involve an inability to understand what one is seeing. Of course, this might occur in some situations, but it is not the norm: when viewing the Carina nebula I might be fully aware of its composition, size, and so on. In short, the spasm of cognitive failure that is central to the standard story has no analogous role in IPP.[25]

So my second claim is apt to be contentious: why endorse it? Here we can point to some cases where, in trying to describe the aesthetic appeal of IPP, people have reached for the term "sublime". In a recent talk about the image of the black hole in M87, the historian Peter Galison spoke about the general appeal of images that "show the unseeable" across the experimental tradition, remarking that '[These images are] sublime in the sense of giving a kind of dignity and broadness to vision beyond what we can [normally] have'.[26] Another example can be found in Robert Crease's (2004) book *The Prism and the Pendulum*, where he discusses the "ten most beautiful experiments" in science. It's noteworthy that when he comes to Foucault's pendulum, a device which allows us to actually see the earth's rotation, he chooses to describe this particular experiment in terms of the sublime. These are two cases where people intimately familiar with the tradition of IPP in the experimental sciences have deliberately chosen to characterise it in terms of the sublime.

But, in defending the second claim, we need not lay too much weight on such examples: we need only examine for ourselves the available options for describing the aesthetic dimension of IPP at issue. The main alternatives to "sublime" appear to be terms like "wondrous" and "awe-inspiring", and indeed some have recently recommended these descriptions in application to science and our experience of natural phenomena.[27] However, it seems to me that they suffer from two defects. First, these concepts remain suggestive of some kind of cognitive failure – what is wondrous or awe-inspiring is, typically, that which defies or stifles understanding. But, again, IPP need not involve cognitive failure. These terms therefore have the wrong ring, laying the emphasis in the wrong place.

Second, and more importantly, these labels fail to clearly point up, or illuminate, the *aesthetic* dimension of the experiences at issue, and their connection to the wider tradition of aesthetic discourse. Talk of "wonder" makes it easy to dismiss IPP as a merely intellectual, rather than properly aesthetic, experience. This is a significant point since the aesthetic dimensions of science and the parts of nature it reveals have too often seemed invisible for want of an adequate aesthetic vocabulary with which to describe them.[28] "Sublime", being the more aesthetically resonant term, offers the most apt way of capturing the aesthetic appeal of IPP that we want to capture in this context (thus, Galison's and Crease's considered choice of the term was, in my view, exactly right).

Should my second claim be accepted, however, we would then need an alternative to the standard account and its insistence on cognitive failure. We can find one by returning to the notion that Sircello saw as lying at the heart of the sublime: the theme of "epistemological transcendence". Sircello follows philosophical tradition in interpreting this in terms of cognitive *failure*: we feel that something transcends our cognitive powers. But this theme might also be glossed in terms of cognitive *success*. That is, we might understand sublime experience as involving awareness of a qualitative expansion of our epistemic access to reality (or alternatively, of our transcending a significant epistemic boundary). Call this the *epistemic expansion* account of the sublime. When we look at celestial objects, such as the Milky Way or the Andromeda galaxy, we may be struck with the feeling that we are taking in an entire "tier" of reality from which we had previously been excluded. This feeling of "expanded access" matches to the oft-noted "positive valence" of sublime experience: it is gratifying, invigorating, even thrilling, to experience such an "epistemic shift".[29]

Defenders of the standard account may be unmoved by such considerations, however. They might respond that to give the word this meaning, and apply it to scientific contexts such as IPP, is simply to, as one writer puts it, "misuse the term" Szécsényi (2021, 54–55). But in this context, what would it mean to "misuse the term"? This might mean employing the term in a way that breaks radically from its historically established usage. Thus, one might argue that the term "sublime" has, historically, chiefly been used to track cases where cognitive failure *does* occur: the traditional 18th-century examples of overwhelming nature such as the roaring cataract, the raging storm, and the looming arctic peak. This response, however, is at odds with the facts, since the use of the term "sublime" has been much broader and more varied than this response suggests.

Even in 18th-century sources, we see a quite different use of the term, for instance in the example cited earlier from Wordsworth: here the idea is that the sublime involves, not cognitive failure and self-reflection, but on the contrary, cognitive success: a transcendence of the limits of normal modes of experience, and contact with something that lies beyond them. When we describe our contact with the imperceptible parts of nature, facilitated by scientific instruments, as "sublime", in the way Galison and Crease do, we are using that term in much the same sense of "transcendence" in which Wordsworth used it. In fact, the term was applied by other 18th-century writers to describe experience of astronomical phenomena in just this way: one poet describes the observer of celestial motions as 'rais'd sublime on Contemplation's Wings', while another describes how 'Newton th' All wise Creator's Works explores/ Sublimely, on the Wings of Knowledge, soars'.[30] Here the sublime is the companion of cognitive success rather than cognitive failure. So to apply the term "sublime" to IPP is not to misuse it, or break radically from historically established use, but only to fail to use it in the way that the *philosophical tradition* has preferred (or rather insisted) it be used.[31]

No doubt, empirically-minded philosophers have favoured the standard account due to the worry that talk of transcendence must end, as it did for Wordsworth, in the capital M mistakes of Mysticism and Metaphysics. However, there is nothing mystical or metaphysical about the transcendence involved in IPP; here we have, as it were, a *naturalised transcendence*, a movement beyond our normal perceptual limitations to experience new aspects of the world. So in this context, an insistence that the standard story captures "the correct use" of the sublime would be quite unjustified on general empiricist grounds.

In fact, one might argue that the epistemic expansion account is not only a viable alternative to the standard story but is in fact superior to it. For it is unclear to what extent the latter works, even for the traditional 18th-century examples of sublime natural phenomena: roaring cataracts, the starry heavens, and the like. Consider, for example, the account of the latter offered by Arcangeli and Dokic. They hold that phenomena such as the Milky Way can trigger a radical limit experience. Because the subject's faculties are felt to be unable to grasp the object, the subject has the feeling that they are "wholly disappearing from the world", this feeling of dislocation or disappearance of the self-representing the negative side of the sublime experience. According to Arcangeli and Dokic, the subject then overcomes this initial feeling of cognitive failure 'by bringing in a sophisticated concept of infinity, which enables her to make sense of what she is confronted with' (ibid.). This allows the subject to feel part of the world once more and represents the positive, pleasurable, aspect of sublime experience.

However, there is a problem with this psychological description because while the Milky Way is vast its size is certainly not incomprehensible (nor it is unperceivable since one can clearly see it in the night sky). On the contrary, we have a good idea of how big it is. Indeed, even if our galaxy were infinite (which it certainly is not), it would still not be incomprehensible, since most adults have a working understanding of the concept of infinity simply from using the natural numbers. So although the starry heavens are vast, it seems that we are never actually in a position where we have no cognitive resources sufficient to grasp them.

In order to find "total cognitive failure" in such an experience, then, Arcangeli and Dokic must hold that we momentarily forget about these cognitive resources, perhaps as we are "blown away" by the immediate appearance of these vast entities, only later remembering that we actually *can* comprehend them.[32] While such temporary amnesia and cognitive paralysis can no doubt occur, there seems no reason, beyond a commitment to the standard account, to think that such temporary amnesia must occur for us to experience the sublime. Imagine observing stars and deep sky objects through a telescope, perhaps as an astronomer explains the origins, sizes of, and distances between, these various objects. Say that during this experience, at no point do you temporarily forget everything you've just been told and experience a temporary global cognitive failure that produces "epistemic angst" or a feeling of "disappearing from the world". Why would this make your experience any less sublime? On the contrary, it seems that understanding the nature and structure of celestial objects renders one's experience more sublime than it would be otherwise by revealing the extent to which one's perception has been "ennobled" (to echo Galison's phrase).[33]

All this being said, we need not go so far as rejecting the standard story altogether. Perhaps the kind of transient psychic breakdown that it describes still has a place in contemporary aesthetic experience. But what is clear is that there is no compelling reason to see this standard story as the only, or even the best, interpretation of the sublime. Should a die-hard defender of the standard account continue to deny the term's applicability to IPP, insisting that the standard story is "the correct account", we can further point out that this position is at odds with the motivation behind the whole revival of the sublime in the first place. For that motivation was that we need the term in order to capture certain experiences of the natural world that remain aesthetically valuable for us today but for which we have no satisfactory aesthetic vocabulary. IPP, however, is exactly this sort of experience. By this reasoning, we ought, then, to use the term "sublime" to capture it, and understanding the sublime as epistemic expansion allows us to do that useful bit of philosophical work.

7.5 Implications: Aesthetic Language and the Aesthetics of Experiment

So far, I have focused on how we can capture IPP's aesthetic dimension by shifting our thinking about the sublime away from the standard philosophical account. To conclude I want to explore two more general issues that this strategy raises: the philosopher's relation to aesthetic language and the way we approach the aesthetics of experiment as a domain of philosophical inquiry.

Regarding the first issue, the strategy proposed here represents a somewhat revisionist approach to aesthetic language. Though I have pointed to some suggestive uses, my chief claim is not that the term "sublime" is widely used to describe IPP, but rather that we can, and ought to, use it in this way. Perhaps it is more common for philosophical treatments, in aesthetics at least, to take up already well-established uses, rather than recommend new ones in this way (according to legend, Frank Sibley drew up his famous list of "aesthetic terms" simply by trawling through volumes of literary criticism in the library). Such hesitancy about trying to shift aesthetic language may be motivated by the thought that art critics are much better situated than philosophers to decide which particular terms are best suited to describe the aesthetic experience of this or that work. This quite sensible humility produces a familiar division of labour: critics deploy aesthetic language and philosophers analyse their usage.

In the case of scientific phenomena such as IPP, however, philosophers have no comparable body of critical writing to rely on in sorting out the best usage. Attempts by scientists and science writers to describe the aesthetic dimensions of their vocation have generally been, as mentioned earlier, infrequent and idiosyncratic.[34] This context therefore calls for a more active role for philosophy in recommending particular uses. And as pointed out above, the revision called for here is not, after all, so radical. Although philosophers have shown a tenacious attachment to the standard story, the term "sublime" has always been an unruly one with a remarkable elasticity, and the epistemic expansion account fits well with the term's most common meaning of elevated or lofty.

The second issue mentioned above concerns one of the ways in which redeploying the sublime to apply to IPP can be of benefit to aesthetic discourse (and hence, ultimately, to aesthetic experience). For not only does it help bring to light an aesthetic aspect of certain experimental practices; it can also broaden the general ways in which we frame the issues of aesthetics and experiment. As mentioned at the outset, discussions of aesthetics and science usually focus on the relationship of aesthetic factors to the traditional epistemic aims of science, such as accurate prediction, understanding, explanation, and the like. The main question asked, whether discussing theories or experiments, is usually: "How do these aesthetic factors contribute to epistemic success?" But the "sublime aspect" of IPP does not fit so neatly into this way of framing the discussion. For if we ask the typical question – "To what extent do practices facilitating sublime experience (i.e., IPP) contribute to empirical success?" – the answer, I suspect, will be, at best, highly context-dependent and historically variable.

At various points in the history of science, the use of various kinds of IPP has certainly helped advance science's epistemic aims, with the spectacular success of the Event Horizon Telescope's image of the black hole in M87 being just one recent example. On the other hand, however, various forms of IPP have often been retired from active duty, so to speak, in favour of more effective methods for hypothesis testing and data collection. The cloud chamber, for instance, was later replaced by the bubble chamber.[35] While it may be plausible to suggest that elegance or simplicity in an experiment is tied, either conceptually or as a matter of empirical fact, to epistemic success, similar claims about the use of IPP are not.

This might seem to suggest that the sublimity of IPP should be viewed as a merely incidental aesthetic feature of experimental science, along the same lines as the pleasing colours of chemical solutions used in the lab, or the pleasant sounds that happen to be generated by some piece of equipment. Though one can find something of aesthetic interest here, these have nothing to do with the operation of science *per se*. In the case of IPP, however, it would be mistaken to conclude that its sublime aspect has no substantial relevance to science and its epistemic aims. Rather, we should broaden our notion of what counts as a "scientific aim" beyond the traditional epistemic goals of accurate prediction, understanding, explanation, and so forth. Science also has more pragmatic and socially-oriented aims, such as communicating scientific knowledge to the public, and cultivating a positive attitude towards, and engagement with, science. In this context, IPP surely has tremendous importance in engaging wider audiences with scientific knowledge and practice. Brightly coloured solutions and gleaming machines are often used in advertising to make scientific work seem attractive and important. But the sublime aspect of IPP can engage the public in a deeper way by allowing the public to experience science's revelation of new parts of the natural world.

For this reason, various forms of IPP (the cloud chamber, for instance) have enjoyed an afterlife of sorts in science education and public demonstration. Astrophotography represents the pre-eminent form of IPP in this regard, having played a major role in shaping public perception of science in recent decades. Astrophotography has been used not only by amateur astronomers, but also by scientific institutions to create images whose primary use is not scientific, but as a means of furnishing engaging aesthetic experiences.[36] Elisabeth Kessler has documented the ways in which the Hubble Heritage Project explicitly drew on the sublime iconography of 19th-century landscape painting in creating aesthetically compelling images for public consumption. Commenting on the allure of these images, Kessler observes that 'they promise a view of something more elusive, a fuller sense of the cosmos' (2012, 171). Commenting on these images and the recent first public pictures released by the James Webb Telescope, Laurence Krauss also explains their appeal in virtue of the fact 'each image represents a new window on an otherwise hidden universe' (2022).

In short, astrophotography, the bubble chamber, and other forms of experiment used in "public demonstration" may not play a day-to-day role in data collection and hypothesis testing, the "bread and butter" epistemic work of experimental science that has long been of central interest to philosophers. But through facilitating sublime experience, they may nonetheless play a critical role in engaging the public with the scientific enterprise. Indeed, in an era when misinformation and distrust of scientific institutions is a significant, if not existential, problem for democratic societies, meaningful public engagement with scientific knowledge and institutions is more important than ever before.

Researchers studying public engagement with science have noted and attempted to explore, the role of aesthetic factors.[37] Though philosophical aesthetics has largely ignored these efforts, analyses of particular aesthetic concepts and experiences, such as the analysis of the sublime proposed here, could be a valuable contribution to them. Philosophers thus have good reason to broaden the scope of "the aesthetics of experiment" as a domain of inquiry, considering both a wide range of aesthetic experiences and a wide range of goals and aims to which these experiences may relate.

Acknowledgements

Thanks to the editors, as well as participants at the September 2021 Aesthetic Nature of Scientific Experiments workshop, for helpful comments and suggestions. This chapter is dedicated to Alex Rueger, teacher and friend.

Notes

1 Recent surveys of philosophical work on aesthetics and science are McAllister (2002) and Ivanova (2017).
2 The emphasis on 'classical' aesthetic features is noted by Kivy (1991). For recent discussions of aesthetics and experiment, see Ivanova (2021) and Ivanova (this volume).
3 In this volume, other 'non-classical' dimensions of the aesthetics of experiment are explored by Murphy, Ivanova, and Ritson.
4 On Greek and medieval attitudes towards landscape, useful sources are Clark (1949) and Glacken (1967). Nicolson (1959) argues that prior to the late 17th century, writers treated mountains with a standardized set of tropes from classical poetry; in actual experience; however, mountains were met with revulsion. For a recent reappraisal of Nicolson's influential account, see Hollis (2019).
5 On the development of landscape painting, the classic source is Clark (1949). Broad historical accounts of the development of the sublime in modern Europe are Monk (1935) and Nicolson (1959). The latter attributes the new appeal of the sublime to a complex series of shifts in European science and theology at the time.
6 On the inadequacy of 'beauty' in this historical context, see Tatarkiewicz (1972) and Stolnitz (1961). In keeping with common practice, my use of

'aesthetics' here is anachronistic, since this term did not come into wide use until the 19th century; for a succinct overview of the conceptual and linguistic issues here see the introduction to Costelloe (2013).

7 Collections of primary texts on the sublime are Ashfield and de Bolla (1996) and Clewis (2019a); the latter contains a concise discussion of the *Peri Hupsos* and its textual history.

8 On the Romantic handling of this and other aesthetic notions, see Gorodeisky (2016).

9 A broad survey of philosophical treatments of the sublime in the early modern period can be found in Brady (2013).

10 Kant distinguishes between the mathematical sublime, produced by vast objects, and the dynamical sublime, produced by objects of great power. The implication that natural objects themselves fail to be sublime on Kant's theory is reassessed by Brady (2013).

11 On the sublime's declining fortunes among the Victorians, see Colley (2016).

12 For a similar worry, see Brady (2013).

13 For more recent defences of the contemporary relevance of sublime experience, see Shapshay (2013, 185–186); Shapshay (2019); and Brady (2013, 185f), as well as the essays in Costelloe (2012).

14 Sircello calls this interpretation of the theme 'ontological transcendence', and points out that 'it is impossible to have any kind of experience at all that presents an object or realm that is in no way epistemologically accessible.' (1993, 547).

15 Along similar lines, Hepburn emphasized the importance of a 'very rapid check to perceptual and imaginative synthesis' (1988, 146).

16 Earlier works outlining an approach along Kantian lines are Crowther (1991) and Hepburn (1988).

17 Keltner and Haidt, while taking note of the sublime as a historically significant category, did not include it in their taxonomy of types of awe. For more on the possible relations between awe and the sublime, see Arcangeli et al. (2020). Accounts along similar lines are given by Clewis (2019b) and Shapshay (2021).

18 See also Arcangeli et al. (2019). The importance of self-negation in the sublime is also stressed by Cochrane (2012).

19 It is worth noting that, in this case, the technology allows us to perceive, via a digital representation, something that could not be perceived otherwise, since no matter how close we got to a nebula, it would be impossible to visualise it due to the distance over which it is spread; on this point, and the nature of astrophotography in general, see Chadwick (2017).

20 For a brief description, see Castelvecchi (2019). Further information on the M87* image can be found at www.eventhorizontelescope.org.

21 This claim assumes some form of scientific realism regarding so-called 'theoretical entities' like electrons. Some anti-realists would deny that many of the experiences I call IPP should be considered instances of 'perception' at all (see, e.g., Van Fraassen (1980)), but I will set such views aside here.

22 This view traces its lineage back to the 18th century and its key notion of disinterestedness; for contemporary defences of views along these lines, see Stecker (2006) and Budd (1995).

23 More precisely, in experiences of IPP, there can be aesthetic value both in what is perceived and the means of perceiving it. Thus, in SEM images, for instance, we can find particular features of the microscopic world itself (the structure of the Drosophila eye, for instance) aesthetically valuable and also find the peering into the microworld itself aesthetically valuable, if we find both valuable to experience for their own sake.

24 Some artists have made use of techniques for IPP in creating Art; on the artistic use of SEM images, for instance, see Tyurina (2020).
25 In another essay, Arcangeli and Dokic do apply their notion of the sublime to science, but here again cognitive failure is central; they argue that it can contribute to the epistemic aims of science by '[drawing] the scientist's attention to highly challenging phenomena and domains of enquiry' (2020, 119).
26 Quoted in Galison and Kessler (2019).
27 Dismissing the sublime as 'aesthetically extinct', literary scholar Philip Fisher instead recommends wonder as a more fruitful concept for contemporary aesthetics (2004, 2). Wonder and awe of course have a long and varied history in science and natural philosophy; see Daston and Park (1998).
28 The persistent tendency to oppose the cognitive and the aesthetic, or even to see them as mutually exclusive, has been highly influential in recent discussions of the aesthetic appreciation of nature; consider Arnold Berleant's remark that 'the concept of "cognitive aesthetics" is actually an oxymoron... because aesthetic values are grounded in perception and cognitive ones are conceptual, both entirely different...' (2016). For a discussion of how this traditional opposition relates to the history of aesthetic perception of experiment in particular, see Rueger (1997).
29 Note that this account requires that the shift be *qualitative*: if I climb a hill and see a part of the landscape that I've never seen before, my epistemic horizon has expanded, as it were, but if this is a qualitative change my experience will not merit description as 'sublime'. Certainly, what counts as a qualitative shift will be importantly dependent on the particular observer. As the Alps lost their dramatic appeal to the Victorians, so the astronomer who is intimately familiar with the night sky may lose a sense of its sublimity (though such a feeling might of course be regained as well).
30 These examples, from John Hughes and Jane Brereton respectively, are taken from Janowitz (2004). She remarks that 'the sublime for Hughes here is a literal elevation, made possible by the grandeur of thought involved in making sense of the universe' (12). Janowitz discusses the eclipsing of this 'benign' sublime by the Romantic or 'unruly' sublime, which emphasized unintelligibility and which became the focus of later writers such as Burke.
31 Is the difference between the standard account and the epistemic expansion account 'merely semantic'? One might argue that on both, our experience is the same: facing a cognitive limit and then ultimately pushing past it. The important difference, however, is that on the standard account, this experience must involve a feeling of panic, disorientation, or arrest, whereas the epistemic expansion account requires no such episode. Another way to put the difference is that on the expansion account, it is a limit that it transcended, whereas on the standard account, our cognitive powers are transcended.
32 They are careful to stress that what is involved in the sublime is only a *feeling* – a metacognitive awareness – of global cognitive failure.
33 It might be thought that by dropping the idea of cognitive failure, the epistemic expansion account fails to do justice to the negative aspect of the sublime, and so won't be able to accommodate some of the traditional examples of sublime phenomena–the roaring cataract, the threatening storm, dangerous animals, and the like. But these cases can also be analysed in terms of epistemic expansion. As Burke noted, the threatening and the dangerous are typically unknown and obscure (because of their danger). Hence to approach and experience these are often to expand one's epistemic range (consider the powerful allure many find in being close to tornadoes and the like).
34 For recent efforts along these lines that address the sublime, see Gross (2018) and Hoffman and White (2011).

35 Rueger (2002) describes the aesthetic appeal of 'mimetic experiments' in the 18th century, suggesting that this appeal endured even after such experiments were dropped from scientific practice.
36 On aesthetic issues in relation to astrophotography, see Chadwick (2017).
37 For a recent treatment of aesthetic dimensions of science education, see Winckman et al. (2021). As in philosophical theorising about the sublime, awe has often been a focus of recent empirical work on science engagement; see, e.g., McPhetres (2019).

Bibliography

Arcangeli, M. and Dokic, J. (2020) 'A Plea for the Sublime in Science', in M. Ivanova (ed.) *The Aesthetics of Science: Beauty, Imagination and Understanding.* New York: Routledge, pp. 104–124.
Arcangeli, M. and Dokic, J. (2021) 'At the Limits: What Drives Experiences of the Sublime', *British Journal of Aesthetics*, 61, 145–161.
Arcangeli, M., Dokic, J. and Sperduti, M. (2019) 'The Beautiful, the Sublime, and the Self', in F. Cova and S. Rehault (eds.) *Advances in Experimental Philosophy of Aesthetics.* London: Bloomsbury, pp. 175–198.
Arcangeli, M., Sperduti, M., Jacquot, A., Piolino, P. and Dokic, J. (2020) 'Awe and the Experience of the Sublime: A Complex Relationship', *Frontiers in Psychology*, 11, 1340.
Ashfield, A. and de Bolla, P. (eds.) (1996) *The Sublime: A Reader in British Eighteenth-Century Aesthetic Theory.* Cambridge: Cambridge University Press.
Berleant, A. (2016) 'Some Questions for Ecological Aesthetics', *Environmental Philosophy*, 13, 123–135.
Brady, E. (2013) *The Sublime in Modern Philosophy: Aesthetics, Ethics, and Nature.* Cambridge: Cambridge University Press.
Budd, M. (1995) *Values of Art: Pictures, Poetry and Music.* London: Penguin Books.
Burke, E. (1757/1968) *A Philosophical Enquiry into the Origin of Our Ideas of the Sublime and Beautiful.* J. T. Boulton (ed.). Notre Dame, IN: University of Notre Dame Press.
Carrit, E. F. (1904) 'The Sublime', *Mind*, 19, 356–372.
Castelvecchi, D. (2019) 'Black Hole Pictured for First Time – In Spectacular Detail', *Nature*, 568, 284–285.
Chadwick, S. (2017), 'Imagination in the Stars: The Role of the Imagination in Artistic Astronomical Photography', *Contemporary Aesthetics*, 15. Available online: https://digitalcommons.risd.edu/cgi/viewcontent.cgi?article=1373&context=liberalarts_contempaesthetics (accessed May 1 2022).
Clark, K. (1949) *Landscape into Art.* London: John Murray.
Clewis, R. (2019b) 'Towards a Theory of the Sublime and Aesthetic Awe', in R. Clewis (ed.) *The Sublime Reader.* London: Bloomsbury, pp. 340–354.
Clewis, R. (ed.) (2019a) *The Sublime Reader.* London: Bloomsbury.
Cochrane, T. (2012) 'The Emotional Experience of the Sublime', *Canadian Journal of Philosophy*, 42, 125–148.
Colley, A. (2016) *Victorians in the Mountains: Sinking the Sublime.* New York, NY: Routledge.
Costelloe, T. (2013) *The British Aesthetic Tradition: From Shaftesbury to Wittgenstein.* Cambridge: Cambridge University Press.

Costelloe, T. (ed.) (2012) *The Sublime: From Antiquity to the Present.* Cambridge: Cambridge University Press.

Crease, R. (2004) *The Prism and the Pendulum: The Ten Most Beautiful Experiments in Science.* New York, NY: Random House.

Crowther, P. (1991) *The Kantian Sublime: From Morality to Art.* Oxford: Oxford University Press.

Daston, L. and Park, K. (1998) *Wonders and the Order of Nature 1150–1750.* Princeton, NJ: Princeton University Press.

Elkins, J. (2011) 'Against the Sublime', in R. Hoffmann and I. B. Whyte (eds.) *Beyond the Finite: The Sublime in Art and Science.* New York, NY: Oxford University Press, pp. 20–42.

Fisher, P. (1998) *Wonder, the Rainbow, and the Aesthetics of Rare Experiences.* Cambridge: Harvard University Press.

Galison, P. and Kessler, E. (2019) 'To See the Unseeable', *Aperture*, 237, 73–77.

Glacken, C. (1967) *Traces on the Rhodian Shore: Nature and Culture in Western Thought from Ancient Times to the End of the Eighteenth Century.* Berkeley, CA: University of California Press.

Gorodeisky, K. (2016), '19th Century Romantic Aesthetics', *Stanford Encylopedia of Aesthetics.* Available online: https://plato.stanford.edu/entries/aesthetics-19th-romantic/#AestNatu (accessed July 1 2022).

Gross, A. (2018) *The Scientific Sublime: Popular Science Unravels the Mysteries of the Universe.* Oxford: Oxford University Press.

Hepburn, R. (1988) 'The Concept of the Sublime: Has It Any Relevance for Philosophy Today?', *Dialectics and Humanism*, 15, 137–155.

Hoffmann, R. and Whyte, I. B. (eds.) (2011) *Beyond the Finite: The Sublime in Art and Science.* New York, NY: Oxford University Press.

Hollis, D. (2019) 'Mountain Gloom and Mountain Glory: The Genealogy of an Idea', *ISLE: Interdisciplinary Studies in Literature and Environment*, 26(2019), 1038–1061.

Ivanova, M. (2017), 'Aesthetic Values in Science', *Philosophy Compass*, 12(10).

Ivanova, M. (2021), 'The Aesthetics of Scientific Experiments', *Philosophy Compass*, 16(3).

Janowitz, A. (2004) 'Adam Smith's Campaign against the Sublime', *The Wordsworth Circle*, 35, 11–16.

Kant, I. (1790/2000) *Critique of the Power of Judgment*, in P. Guyer and E. Matthews (trans). Cambridge: Cambridge University Press.

Keltner, D. and Haidt, J. (2003) 'Approaching Awe, a Moral, Spiritual, and Aesthetic Emotion', *Cognition and Emotion*, 17, 297–314.

Kessler, E. (2012) *Picturing the Cosmos: Hubble Space Telescope Images and the Astronomical Sublime.* Minneapolis, MN: University of Minnesota Press.

Kivy, P. (1991) 'Science and Aesthetic Appreciation', *Midwest Studies in Philosophy*, 16, 180–195.

Krauss, L. (2022), 'Watch This Space', *The Globe and Mail*, July 16.

McAllister, J. (2002) 'Recent Work on Aesthetics of Science', *International Studies in the Philosophy of Science*, 16, 7–11.

McPhetres, J. (2019) 'Oh, the Things You Don't Know: Awe Promotes Awareness of Knowledge Gaps and Science Interest', *Cognition and Emotion*, 33, 1599–1615.

Monk, S. (1935) *The Sublime: A Study of Critical Theories in XVIII-Century England.* New York, NY: Modern Languages Association.

Nicolson, M. H. (1959) *Mountain Gloom and Mountain Glory: The Development of the Aesthetics of the Infinite*. New York, NY: W.W. Norton & Company.
Parsons, G. and Rueger, A. (2000) 'The Epistemic Significance of Appreciating Experiments Aesthetically', *British Journal of Aesthetics*, 40, 407–423.
Rueger, A. (1997) 'Experiments, Nature and Aesthetic Experience in the Eighteenth Century', *British Journal of Aesthetics*, 37, 305–322.
Rueger, A. (2002) 'Aesthetic Appreciation of Experiments: The Case of 18th-Century Mimetic Experiments', *International Studies in the Philosophy of Science*, 16, 49–59.
Shapshay, S. (2013) 'Contemporary Environmental Aesthetics and the Neglect of the Sublime', *British Journal of Aesthetics*, 53, 181–198.
Shapshay, S. (2019) 'A Theory of Sublime Reponses, the Thin and the Thick', in R. Clewis (ed.) *The Sublime Reader*. London: Bloomsbury, pp. 329–339.
Shapshay, S. (2021) 'A Two-Tiered Theory of the Sublime', *British Journal of Aesthetics*, 61, 123–143.
Sircello, G. (1993) 'How Is a Theory of the Sublime Possible?', *Journal of Aesthetics and Art Criticism*, 51, 541–550.
Stecker, R. (2006) 'Aesthetic Experience and Aesthetic Value', *Philosophy Compass*, 1, 1–10.
Stolnitz, J. (1961) '"Beauty": Some Stages in the History of an Idea', *Journal of the History of Ideas*, 22, 185–204.
Szécsényi, E. (2021) 'Remarks upon the Aesthetics of the Night Sky', *Espes*, 10, 51–63.
Tatarkiewicz, W. (1972) 'The Great Theory of Beauty and Its Decline', *Journal of Aesthetics and Art Criticism*, 31, 165–180.
Tyurina, A. (2020) 'Searching for New Aesthetics: Unfolding the Artistic Potential of Images Made by Scanning Electron Microscopy', in R. Earnshaw, S. Liggett, P. Excell and D. Thalmann (eds.) *Technology, Design and the Arts – Opportunities and Challenges*. Cham: Springer, pp. 103–119.
Van Fraassen, B. (1980) *The Scientific Image*. Oxford: Clarendon Press.
Winckman, P., Prain, V. and Tytler, R. (2021) 'Aesthetics, Affect, and Making Meaning in Science Education: An Introduction', *International Journal of Science Education*, 44, 717–734.
Wordsworth, W. (1798/1967), 'Lines: Composed a Few Miles above Tintern Abbey', in D. Perkins (ed.) *English Romantic Writers*. San Diego, CA: Harcourt Brace Jovanovich, pp. 209–211.

8 Aesthetic Cognitivism, Experimental Shielding, and Explanatory Reasoning

James Nguyen[1]

8.1 Introduction

In recent years, philosophers of science, in particular those who focus on scientific models, have drawn, and continue to draw, on the tools and frameworks originally introduced in the context of aesthetics and the philosophy of art. Pertinent examples include: the use of Kendall Walton's (1990) analysis of make-believe to provide either a deflationary ontology of scientific models (Frigg and Nguyen 2016), or an account of how they work more generally (Toon 2010); the use of accounts of aesthetic representation to support and justify accounts of scientific representation ((French 2003) draws on (Budd 1993); (Elgin 2017) and (Frigg and Nguyen 2020) draw on (Goodman 1976)); and discussions concerning the role of aesthetic virtues in scientific theory choice (Ivanova 2017; O'Loughlin and McCallum 2019). As it stands, this flow of ideas from the philosophy of art to the philosophy of science seems to be a productive one (in addition to the aforementioned citations, see, e.g. the collected volumes (Bueno et al. 2017) and (Ivanova and French 2020)).

But the flow doesn't have to be one-way. If framing scientific practice, theories, and models in aesthetic terms is fruitful, we might come to think of science and art as closer cousins than is typically recognised. And if it turns out that the distinction between science and art is so porous – i.e. if it proves valuable to think of aspects of science using tools from art and the philosophy of art – then it may also prove valuable to theorise about art using tools from science and the philosophy of science.

Stated at such a level of generality, this methodological suggestion doesn't amount to much. There are many ways in which we could embark on aesthetic explorations in terms of what scientists, and philosophers of science, have said about science. And some (possibly many) of these cross-discipline migrations would prove fruitless. But there are certain aesthetic questions where we might expect the strategy to pay off. In particular, if we take modern science as an exemplar of human cognitive achievement, then frameworks for analysing this achievement

DOI: 10.4324/9781003263920-9

could, plausibly, help us understand how art provides related successes (if it does at all), i.e. how we might learn from art.

Developing this idea is my task here. To keep things manageable, I focus on the cognitive function, and value, of literary fiction (rather than art in general). There is a clear sense in which we learn from such works, the task is to pinpoint how we do so. In this chapter, I discuss ways in which they play this function by drawing on two analogies with scientific methodology (and philosophical reflection thereon). First, works of literary fiction can describe appropriately constrained systems, where an explanatory factor can be observed to act in isolation. Through this, we can come better understand the contribution this factor makes in actual systems. In this sense, the cognitive value of such works should be understood as analogous to the cognitive value of (particularly, thought) experiments.[2] Second, through works of literary fiction, we can investigate how explanations that we suspect hold in the actual world fare in the world of fictions. So when faced with competing hypotheses about what explains some set of observations in the actual world, learning that one of them (and not the others) acts as an explanation in an appropriate world of literary fiction can serve to confirm that hypothesis at the expense of its competitors. In this sense, the cognitive value of such works should be understood as analogous to the cognitive value of modelling strategies that employ "robustness analyses" (particularly as analysed by Schupbach 2018).[3]

This chapter is structured as follows. I begin by outlining the different ways in which works of literary fiction have been thought to hold cognitive value. In particular, I focus on the idea that when we learn from literary fiction, we, ultimately at least, are learning about the actual world. This serves to precisify the type of cognitive value I'm interested in in this chapter (Section 8.2.1). I then change tack and provide a broad sketch of how to conceptualise the cognitive function served by experiments, thought experiments, and robustness analysis (model-based or otherwise) (Section 8.2.2). I tie these discussions together in Section 8.3, where I demonstrate how works of literary fiction can be understood as functioning as thought experiments, and moreover, can confirm certain hypotheses by ruling out others. In Section 8.4, I answer some objections to my account, and in Section 8.5, I situate the project within a broader picture of human cognitive endeavour; cautioning against the idea that it's an attempt to assimilate the cognitive value of art into the scientific domain and advocating instead a more humanistic way of analysing how we use simplified systems to help us understand the complexities of the world we're a part of.

8.2 Rehearsals

In this section, I provide the requisite background to the debates concerning the cognitive value of literary fiction on the one hand, and both experiments and robustness analysis on the other. This serves to precisify the target of this chapter, the sort of cognitive role of literary fiction I'm interested in. I also outline how the practices of presenting a scenario in which some feature of the world can be observed to act in isolation, and demonstrating how an explanation fares compared to its competitors in actual and non-actual scenarios, can serve as cognitive achievements.

8.2.1 Cognitive Value of Literary Fiction

My focus in this chapter is on the cognitive value of "literary fiction": works that are considered both literary and fictional. As Gaut (2005, 2007) emphasises, these terms should not be considered co-extensive; there are plenty of works of literature that are not necessarily fictional (e.g. Anne Frank's *Diary of a Young Girl*), and works of fiction that are not literary (e.g. E. L. James' *Master of the Universe*, an erotic *Twilight* fan fiction later developed into *50 Shades of Grey*). But exploring the cognitive value of (elements in) their intersection poses a particularly pertinent pair of philosophical questions: how can we learn from things that are not put forward as literal descriptions of the actual world (i.e. fictions), and if we do learn from them, how does this relate to their aesthetic value (i.e. their status as works of literature)?

What do I mean by "not put forward as literal descriptions of the actual world"? There are various different ways in which this could be explicated. For example, we could take such work to describe non-actual concrete possible worlds (Lewis 1983), or we could take them to function as prescriptions to imagine in a game of make-believe (Walton 1990). But all that's important for my current purposes is that when reading such work, we do not take them to describe any actual system, and it is of no detriment to such work that they fail to do so.[4]

This does not entail that such works are unconstrained. I will help myself to the idea that claims can be true or false *within the fiction*. In particular, it makes sense to refer to the presence (and absence) of events occurring in "the world" of a literary fiction. Moreover, such events can enter into explanatory relations, in those "worlds". For example, I will take it for granted that claims like "in *Changing Places*, Morris Zapp's desire to be ingratiated with Hillary Swallow explains why he recommends her husband be promoted to Senior Lecturer" are true, and claims like "Morris Zapp's shyness explains why he moves to England" are false. Moreover, that such claims are true (or false) does not require that they (or their negations) be explicitly stated in the fiction.[5] So as I understand them, literary fictions have an internal explanatory structure, but do not describe any actual system in the world.

What about "cognitive value"? In the aesthetics literature, aesthetic cognitivism (hereafter "cognitivism") is often presented as the conjunction of two claims: (i) the *epistemic* claim that works of art hold cognitive value, and (ii) the *aesthetic* claim that this cognitive value contributes to the aesthetic value of the works in question (Gaut 2007, see in particular Chapters 7 and 8).[6] My primary concern in this chapter is the epistemic claim: in what sense, if any, are works of literary fiction cognitively valuable?

Let's narrow down the kind of cognitive value that is my target here. There are many ways of conceptualising the (purported) cognitive value associated with literary fiction. When (if) we learn from such work do we gain new knowledge; do we reconceptualise things we already know; or do we gain some other cognitive good, like understanding, or improved intellectual virtue or skill? And if we do gain new knowledge, what kind of knowledge are we talking about: propositional, practical, experiential, or something else? And for each of these cognitive goods, what are their subjects: do we learn about the fictional world, the actual world, our concepts, morality, or what?[7]

I will argue that works of literary fiction can provide us with propositional knowledge about the actual world.[8] This sets the bar against which the aesthetic cognitivist's epistemic claim is measured relatively high. As a result, I take it that this reduces the importance of the cognitivist's second claim (that what we learn from work can impact its aesthetic value). To explain: it is uncontroversial that we can learn *something* from works of literary fiction. For example, from the physical features of a text (e.g. that it is printed on a particular brand of paper), we learn about its origin (e.g. that it was printed at a particular printing house). But this doesn't impact its aesthetic value (at least in any obvious way). So as a result, anti-cognitivists can grant that this kind of cognitive value may be associated with works of art, but rule it unimportant insofar as it fails to satisfy the aesthetic claim. It is also uncontroversial that we can learn from features of the work itself (e.g. that the characters in the work speak a particular dialect), something about the context in which it is written (e.g. that the author lived in a particular region in which the dialect is spoken). And so again, it is reasonable for the anti-cognitivists to grant that such cognitive contributions may be had, but to further argue that they fail to impact the aesthetic value of the works in question.[9] This suggests that for some cognitive good x, the easier it is to demonstrate that works of literary fiction deliver x, the more important (and the harder) it is to demonstrate that delivering x contributes to their aesthetic value. As a result, if we can show that works of literary fiction deliver cognitive goods in a particularly surprising manner – if they hold kinds of cognitive value that we would not expect them to hold (given their status as fictions) – then the less important it is to demonstrate that this plays an aesthetic role (or at the very least: it is suitable to defer answering this question in the first instance).

It is difficult to see how we can gain propositional knowledge about the actual world from works of literary fiction, since: (i) we don't take them to describe, or to provide access to in some other way (at least in the first instance), any actual system, and (ii) propositional knowledge about some system (in contrast to other cognitive goods, like cognitive skills or capacities) requires a description of, or access to, that system. So how can we be expected to gain such knowledge from such works? The mechanisms that I describe in this chapter will demonstrate precisely how this works. As a result, I take it that providing evidence for the aesthetic cognitivist's epistemic claim, where "cognitive value" is understood in this sense, is sufficient to make a positive contribution to the debate, even without the additional claim that this thereby contributes to the aesthetic value of the work (although I briefly return to this question in Section 8.4).

8.2.2 *Styles of Scientific Inference*

I now introduce the two aspects of scientific practice I will draw upon to understand the cognitive value of literary fiction: experiments and robustness analysis. In light of the above discussion, what's important about these practices is precisely that they allow us to learn about some (actual) system or explanatory factor throughout the investigation of another (sometimes non-actual) system.

8.2.2.1 Experiment

Let's start with physical experiments. Suppose you want to measure the value of g, acceleration due to the earth's gravity, at some position on the earth's surface. To do so, you take a pendulum of some known length to that location, set it oscillating (at a relatively low amplitude), and time how long it takes to tick (and to tock). Using this, in combination with some assumptions from classical mechanics, you can infer from (e.g.) the fact that a just shy of one-meter long pendulum has a close to two second period, that (e.g.) the value of g at that location takes a value around 9.8 m/s^2.

How does this work? Well, one thing that we know is that the period of a pendulum is affected by a variety of forces beyond the earth's gravity: there is friction at the pivot, air resistance acting on the bob and the cord, and there are additional gravitational forces associated with bodies beyond the earth. So if you want your experiment to measure g in particular, then you need to try to *shield* your measurement device from these additional forces. You would, for example, grease the pivot (minimising friction), ensure that the cord of your pendulum is thin and your bob relatively small (minimising air resistance), and so on. This is a particular example of a general theme. Experiments need to be set-up carefully: depending on the details of the experimental apparatus, you need to isolate the feature of interest (in this case, the effect of gravity) and control the way in which your measurement device (in this case, your pendulum of a known length) interacts with it. And the more confident you are that you have shielded your measurement device from additional forces, the more confident you can be in the result of your experiment (i.e. the closer your calculated value should be to the true value of *g*, at that location).[10]

Shift the example slightly. Suppose you are interested in the relationship between the length of a pendulum and its period. You could proceed by constructing a series of physical experiments: varying the length of the pendulum and investigating the resulting periods. And again, if performing such an (series of) experiment(s), you'd be well-served to ensure that the pendula were shielded from additional forces (in this case, you would at least try to ensure that in varying the length of the pendulum you keep the impact of extraneous forces fixed).

But there is another way to proceed. Rather than constructing a *physical* experiment, you can construct a *thought* experiment – in this context usually called a "model" – instead.[11] Rather than attempting to minimise the extraneous forces acting on a real pendulum, you can consider an imaginary pendulum, which, in the world of the thought experiment, is acted on by gravity alone. In this thought experiment the "ideal" pendulum is an entirely isolated system consisting of a point mass suspended on a massless cord from a frictionless pivot (on which no forces other than earth's gravity act). This system's equation of motion can then be solved, and how the length of the pendulum impacts its period explicitly calculated.

Entering the world of thought experiment is motivated by reasons analogous to those operating in the physical experiment: you want to shield the experimental system from any extraneous forces, and "observe" how the features of interest act in isolation.[12] Whilst, with some effort and experimental ingenuity, this can be done imperfectly in the case of physical experiments, thought experiments offer scientists the opportunity to do this completely: it is only by entering the world of thought that we can guarantee that a pendulum is acted on by gravity in absolute isolation. This is (one of) the cognitive benefit(s) of constructing thought experiments.

Now, notice that the results of either kind of experiment, either physical or in thought, do not remain restricted to the immediate system in which they are explored. When measuring the value of *g* at a particular location, your interest (presumably) went beyond gravity at that location: you could infer that *g* was (roughly) this across the surface of the earth. And when performing a thought experiment to determine the relationship between pendulum length and period, your interest (presumably) went beyond the ideal pendulum: you can infer from what you have found in the thought experiment to the (imprecise) relationship between the length and period of actual pendula (and explain, for example, how adjusting the length of the cord would impact the period of the clock in the Vice Chancellor's office).

These inferences, from experimental set-ups to a broader set of (actual) systems, are not *guaranteed* to deliver true conclusions. In the cases at hand, increasing the length of the pendulum will in fact increase the air resistance it faces, which will (slightly) impact the length/period relationship of actual pendula. And since the value of g at a location depends on that location's distance from the centre of the earth, it will vary across the earth's surface, and therefore typically won't exactly match any given measurement at a particular location.[13] So the success of the (physical or thought) experiment *alone* does not guarantee that we can make successful inferences about some broader class of systems. Nevertheless, there is a clear sense in which such experiments *contribute* to our knowledge of systems beyond themselves (I return to this point in Section 8.4).

To summarise then: experiments work by shielding features of interest, allowing us to observe how they act in isolation. One way to do this is to construct thought experiments, according to which those features are isolated by design. Then, from our investigations into the behaviour of these isolated features, we can thereby come to know something about how they act in broader classes of systems (where they may act in combination with additional features).

8.2.2.2 Robustness Analysis

Some experiments and modelling exercises work by allowing us to explore the behaviour of features of interest in (relative) isolation. But this does not exhaust the function of experiment or modelling. In some instances scientists experiment, and model, in order to explore whether some hypothesis holds in *multiple* different scenarios: to test whether a hypothesis is *robust* across changes to an experimental, or modelling, set-up.

We can illustrate this with an example from (Schupbach 2018). In 1827, botanist Robert Brown observed that pollen particles suspended in water moved in a particular way. He ruled out that the motion was explained by currents or evaporation of the water, by testing whether it continued to occur in still water, and at different temperatures, and hypothesised that it was due to the particular shape of the pollen he was observing. However, this was ruled out when he observed the same motion in pollen particles of different plants. He then tested a number of different potential explanations for the motion (e.g. sexual drive inherent in pollen, or vital force in organic material) and continued to find the motion. Additional experiments by later physicists ruled out further hypotheses, revealing that the motion persisted in spite of various changes to the experimental set-up (e.g. using different containers and mediums, and so on). 'Eventually, almost eighty years after Brown first puzzled over the movements of his pollen granules, Einstein's 1905 work on molecular theory and thermodynamics provided the missing link, making sense of Brownian motion' (Schupbach 2018, 276).

So we observe some type of behaviour (Brownian motion) in a number of different instances, i.e. in a number of different experimental setups. By varying the setups, we rule out various hypotheses concerning what explains this behaviour (or what explains this particular instance of this behaviour). Eventually, once we had the hypothesis offered by Einstein's work (including the hypotheses that thermodynamics and molecular theory are applicable to each experimental system we have observed the behaviour of, and would cause Brownian motion were they true), by ruling out the competing hypotheses, we confirm the actual explanation of the motion.

Robustness analysis is not restricted to multiple physical measurements. Another form of robustness analysis occurs in the context of model-based science. Here the different scenarios under which the hypothesis holds do not concern different measurement contexts but are different model systems. An example familiar to philosophers of science is offered by Weisberg and Reisman's (2008) discussion of the so-called Volterra Principle: 'ceteris paribus, if a two-species, predator-prey system is negatively coupled, then a general biocide will increase the abundance of the prey and decrease the abundance of the predators' (p. 114).

Let's see how robustness analysis works in this case. Volterra (1926) observed a change in the proportion of different fish (i.e. sharks, predators, and food fish, prey) available at the fish markets during and after the First World War. He reasoned that the increased fishing after the war was a biocide – an intervention that decreases the size of the two populations uniformly. As a result, he wanted to explore how such biocides would impact the relative sizes of such populations. He constructed a model of a predator-prey system by specifying two differential equations that together describe how the size of the predator or prey population develops as functions on one another. According to these equations, the two populations are negatively coupled: increasing the abundance of the predator population decreases the abundance of the prey population and increasing the abundance of the prey population increases the abundance of the predator population.

Comparing a solution to these equations that represents the pre-biocide system with one representing a post-biocide system, Volterra found that the prey-predator ratio is higher in the latter. So: in the model (like the world) a general biocide increases the relative abundance of the prey to the predators. Call this general feature the "Volterra Property". We then ask, what explains this property? Volterra hypothesised that it is explained by the negative coupling of the populations in question, in both the model and more generally, i.e. he proposed the Volterra Principle.

However, one might worry that the fact that the model exhibits the Volterra property is an artefact of seemingly irrelevant details of the model (e.g. the value of the coefficients in the differential equations, the fact that it is defined on the reals, or other idealisation assumptions), rather than the negative coupling. Considering the model in isolation, we don't necessarily have a good reason to believe that it's the negative coupling that explains the effect of the biocide, even in the model system. And if we're not confident in this explanatory relationship in the model, then why should we be confident in the relationship general?

Robustness analysis provides a way of addressing this concern: we can *test* whether the Volterra property depends on negative coupling or the irrelevant details of the modelling by holding the former fixed, varying the latter, and seeing whether the Volterra property remains. This can be done in various different ways. First, we can consider "parameter robustness": does the result vary if the parameters in the differential equations are changed? In this case, the answer is no. Second, we can consider "structural robustness": does the result vary if we alter the structure of the equations themselves? Again, the answer is no (see Weisberg and Reisman 2008, who introduce a term representing carrying capacity and observe that the Volterra property still holds). Third, we can consider "representational robustness": does the result vary if we alter the representational framework in which the model is embedded? Again, in this case, the answer is no (Weisberg and Reisman 2008, show that an agent-based model of the prey-predator, a model which explicitly represents interactions between individual predator and prey still exhibits the Volterra property).

The results of these investigations show that despite varying the contexts in which a feature of the model(s) – the negative coupling of the predator-prey population – is embedded, it remains associated with some feature – the Volterra Property – of interest. As a result of this analysis, one is led to be more confident in our initial explanatory hypothesis: it's the negative coupling of the populations that explains the relative impact of a biocide, rather than other details of the individual models (since those other details vary across the model scenarios considered). The final hope then is that this practice, of exploring whether a particular feature (in this case, the fact that negatively coupled populations are associated with the Volterra property) is robust across modelling scenarios, increases our confidence in the general explanatory claim expressed by the Volterra Principle: that negative coupling *in general* explains why biocides favour the prey.

There have been various different attempts to explicate the details of the style(s) of reasoning associated with practices such as the above, but for my current purposes I rely on Schupbach's (2018) discussion.[14] He provides a Bayesian reconstruction of such reasoning that turns on the idea that sequences of experiments/models can be used to discriminate between competing explanatory hypotheses. Four conditions need to be met in order to confirm such a hypothesis via such reasoning: a result R has been detected previously via different means (let our evidence E be the conjunction of these past detections $E = R_1 \wedge R_2 \wedge \ldots \wedge R_{n-1}$) (Past Detections); there are at least two hypotheses H and H', which both purportedly explain E (Success); but they compete with one another with respect to E (Competition); there is an additional explanatory discriminating means for detecting R, such that one hypothesis would explain such a detection (R_n), and the other would explain not detecting R in this manner ($\neg R_n$); and finally the new detection occurs (R_n). If these four conditions are met, then detecting R through the nth means confirms one of the explanatory hypotheses (H say), at the expense of its competitors (H' say). Schupbach demonstrates formally what he informally describes as the 'target hypothesis H soaks up the epistemic virtue that H' had going for it prior to being eliminated; spoils to the victor' (2018, 294).

To see how this works as applied to the above examples, we have: R = Brownian motion, and R_1 = Brownian motion with pollen at room temperature in a rectangle container, R_2 = Brownian motion with pollen at cooler temperatures in a rectangle container, R_3 = Brownian motion with pollen at room temperature in a cylindrical container, etc.; H = molecular theory and thermodynamics, H' = vital force of organic matter; and R_n = Brownian motion with inorganic material (at some temperature in some container). In this case, we have observations of Brownian motion (R) in various circumstances (E), and we have two competing hypotheses for what we have so far observed (H and H'). With respect to another observation, an additional experimental set-up using inorganic matter (R_n or $\neg R_n$, depending on the result of the experiment), one of those hypotheses, i.e. Einstein's, will explain the (potential) resulting Brownian motion in the set-up (R_n) and another, i.e. vital force of organic matter, will explain the (potential) failure to observe Brownian motion in the set-up ($\neg R_n$). Once the experiment with the inorganic matter is performed, and Brownian motion is observed (R_n), this serves to confirm H at the expense of H'.

In the second case, we have: R = biocides favour the prey, and R_1 = biocides favour the prey in Volterra's model with some set of coefficients, R_2 = biocides favour the prey with some other set of coefficients, R_3 biocides favour the prey in the modified model that includes the carrying capacity of the environment, etc.; H = negatively coupling; H' = (e.g.) no agent-level interactions (i.e. population size modelled only at the level of populations rather than individuals); and R_n = biocides favour the prey in an agent-based model (with negative coupling). In this case, we have a sequence of population-level models each exhibiting the phenomena R, i.e. that biocides favour the prey. We may have competing hypotheses for what explains these model phenomena: H negative coupling of the populations in the models; or H' the models being stated at the population level, without reference to the agent interactions. With respect to another model, the agent-based version of Weisberg and Reiman, H would explain R_n, observing that a biocide favours the prey in that model, and H' would explain $\neg R_n$, observing this isn't the case in the model. When the model is investigated, it turns out that the biocide favours the prey, and thus we confirm that H, rather than H', explains R.

In general then, the idea is that given our limited observations in the world (or model) of some tokens of some type of feature, we may be faced with competing explanatory hypotheses that account for them. In such cases, when we observe another token of the same type, if the result is robust across that novel scenario, and if this result is accompanied by one of the potential explanations, but not the other(s), this serves to confirm the former explanation at latter(s)'s expense. In this way, we can increase our confidence in certain explanatory hypotheses by virtue of observing that they hold in a robust way. And in doing so, our observation of how the token behaves in the novel scenario provides us with increased confidence (a cognitive gain) with respect to claims about how it behaves in other systems.

8.3 Learning from Fiction

I have clarified the sort of cognitive value of literary fiction we are interested in – the fact that they can yield propositional knowledge of the actual world – and argued that if this can be shown, then the question of whether such value contributes to their aesthetic value becomes of secondary importance. I have also outlined the structure of two aspects of practice that yield analogous knowledge in the scientific context. It remains to demonstrate how works of literary fiction share this structure. I do so via the use of five examples.

8.3.1 Fictions and Isolation

I first argue that we can learn from literary fictions in exactly the manner presented in Section 8.2.2.1: works of literary fiction describe shielded systems in which we can "observe" how particular features of interest operate isolation, just as the ideal pendulum is an example of a shielded system in which we can "observe" how pendula period and length interact as isolated from features such as air resistance, friction, and so on. And from this, we can gain explanatory knowledge about these features, as they are realised in the actual world. To illustrate this I present two examples of literary fiction where this style of reasoning is on display.

To warm up we can recapitulate Elgin's discussion of how Austen's *Pride and Prejudice* acts as an isolated system exploring social relationships, focusing in particular on the interaction between love, the social role of marriage and family, the subtleties of the British class system, and the difficulty in understanding others and ourselves.[15] By engaging with the development of Elizabeth Bennett in a tightly constrained environment – a couple of handfuls of characters, largely interacting in rural Hertfordshire in 19th-century England – we observe the impact of the expectations, put on us by ourselves and others, coupled with the difficulties in reading the motivations behind certain actions (our own and those with whom we interact) when said actions are embedded in, and contribute to, a social system driven by the interplay between social forces (e.g. considerations to do with expectations and social position) and individual motivations (e.g. awkwardness, love, and other emotions).

The world of the fiction thus acts to display these aspects of human life as shielded from the others that, in the actual world, act in combination with them. For example (as noted by Elgin 2017, 258), *Pride and Prejudice* was written during the Napoleonic Wars, which presumably had an impact on Austen given she had family members who served in the navy. Indeed the wars provide the backdrop against which the story plays out (consider, e.g. the regiment that causes a considerable stir on their arrival to town). And yet this wider geopolitical context plays no significant role in the development of the world of fiction. The same can be said about Elizabeth's reasoning about others, in the context of her attempting to understand their, and her own, feelings. In the contemporary setting, for many of us (I assume, at least in the global academic profession) our thoughts and plans about our romantic and social relationships are intertwined with considerations about where in the world we may live, and how we will make a living doing so. Yet such considerations are alien to the world of *Pride and Prejudice*.[16] The same could be said about any number of other social factors that are excluded from the novel: extreme class differences, differences in culture, language, sexuality (including physical compatibility), or religion, all of which can interact with the sorts of social and personal forces explored in the novel in the course of actual domestic affairs. Thus, the novel constructs a shielded system where we can observe the former factors acting in relative isolation, and by exploring how they play out in the restricted context, we can come to better understand how they work. And from this we can better understand the contribution they make in actual systems, where they are embedded within a broader social nexus, thus providing us with propositional knowledge about the actual world.

Ishiguro's *The Remains of the Day* provides another example of this style of learning from literary fiction. In this case, the constrained system is the perspective of Stevens, a butler in the twilight of his career, reflecting on his life and choices that led up to his current position. Exploring his intertwined professional and personal history from the first-person perspective, we are provided with a demonstration of how one's sense of honour, dignity, and self-perception more generally, can distort how one interprets the situations one finds oneself in. We come to recognise that Stevens' loyalty to his previous employer prevented him from properly recognising the implications of Lord Darlington's efforts for appeasement with Nazi Germany, even when it involves firing two Jewish maids who work under Stevens. And we also recognise how one can fail to appreciate the strength of one's feelings for another, and the life possibilities that could be realised by admitting to them, when Stevens recounts his interactions with Miss Kenton. This lesson is agonisingly reinforced at the end of the novel, when Stevens meets with Miss Kenton, learns that she married another man and, reflects on whether it is better to focus on the present, rather than past opportunities missed.

Such a demonstration of how Stevens' (self-perceived) dignified restraint prevents him from embracing opportunities for happiness, and the manner in which he describes the situation from his perspective (with the reader understanding how his values impact his interpretation of various significant events in his life) again provide a microcosm in which we can observe how this particular feature of human thought acts, as shielded from other features that in actual systems would play a significant role in situations like Stevens' employment under Lord Darlington, and his relationship with Miss Kenton.[17] So again, the work provides us with a tightly constrained system – Stevens' limited perspective, embedded within working at Darlington Hall, focusing particularly on a few particular facets of human thought – we are provided with a demonstration of how a limited set of factors can "play out". And I take it that many of us are (I confess, I am) familiar with this style of thought. But in the actual world, they are combined with any number of additional factors resulting in a complex messy web of interpretation and action. By observing how they act in isolation in Stevens' case, we can thereby come to understand (and, perhaps, address) how they contribute to our actual mental lives.[18] Again, from "observing" the isolated system associated with the work of literary fiction, we can thereby increase our knowledge about how the features in that system contribute in systems in the actual world.

8.3.2 Fictions and Explanatory Robustness Analysis

So far in this section, I haven't argued for anything particularly novel. As mentioned previously, the idea that works of literary fiction act to display how isolated features behave when shielded from features that they interact with in actual systems has been argued forcefully by (Elgin 2017).[19] But I don't think this exhausts their cognitive value. I now argue that we can also learn from literary fictions in exactly the manner presented in Section 8.2.2.2. We can be faced with some set of observations in the actual world that we might plausibly want to explain. And we might have multiple competing hypotheses for what explains these observations. Showing that one of these hypotheses, but not the other(s), is robust in the sense that it serves to explain some similar observations *in a work of fiction*, can confirm that hypothesis at the expense of the other(s). Again, I will demonstrate how this works, this time via three examples. For each, we need to identify the important moving parts required in order for Schupbach's analysis to go through: the some previous observations ($E = R_1 \wedge R_2 \wedge \ldots \wedge R_{n-1}$); the competing explanatory hypotheses (H and H'); and the additional observation in the work of fiction (R_n), which is explained by one hypothesis, and not the other(s).

Return again to *Remains of the Day*, and imagine that you share Stevens' "quintessential Englishness".[20] Suppose that, like Stevens, you are reflecting on your life and (in some sense at least) you come to recognise how your self-perceived "dignity" has coloured your interpretation of past events (in a manner similar to how Stevens interprets his enabling of Lord Darlington's antisemitism and contributions to appeasement) and provided you with an excuse for not taking full responsibility for your choices and actions, even where these have had dramatic consequences for how you have lived your life and your (missed) opportunities.[21] Let R_1 denote this aspect of how your life as turned out. You might then ask: is there an explanation for this, i.e. is there an explanation for R_1? (And if you have recently read *Remains of the Day* you might ask whether this explanation also explains the relevant aspects of Stevens' life, which I will denote by R_2.)

Here, I think, at least two possible answers present themselves. Call the first the "English Hypothesis" (H). According to this hypothesis, there are certain character traits, many of which are exemplified by Stevens in the novel, that are associated with a particular national stereotype – including a slightly pompous self-image; a commitment to the "stiff-upper lip"; and a respect for one's position in society – and that it is precisely by conforming to the details of this stereotype that explains why your and Stevens' lives turned out the way that they have.[22]

An alternative hypothesis eschews an explanation in terms of such stereotypes. According to what we might call the "General Self-Deception Hypothesis" (H') it's neither your, nor Stevens', *Englishness* that performs the explanatory role: it's more generalised versions of the character traits, common across different cultures albeit realised in different ways. These traits include our tendency for self-deception via invoking excuses about what is "expected" or "appropriate" and they explain the missed opportunities and the regret, combined with the disconnect between events as seen from an external perspective and as we (try to) interpret them. But they are not peculiar to any particular culture. So you have two related observations, R_1 and R_2, that you recognise to be tokens of the same type (even though the latter is an aspect of a literary fiction, and the former refers to an aspect of an actual system), and you have two competing hypotheses for these observations: H, according to which they are explained by particularly English character traits, and H', according to which they are explained by traits that are shared more generally. Both hypotheses seem relatively successful at explaining R_1 and R_2 (for the latter, consider a reading of *Remains of the Day* that takes Stevens as an exemplar of the English stereotype in this regard, and interprets the novel through this lens), and moreover they compete with one another: the former focuses on the explanatory contribution of a national stereotype, the latter focuses on the explanatory contribution of general character traits. Following the structure of (Schupbach's (2018) account of) robustness analysis, one way that we can discriminate between these hypotheses is to consider an additional system where the predictions of the two hypotheses will come apart. For example, we could consider a system set in post-Second World War Japan, exploring how a Japanese artist who enabled Imperial Japan, in part through their contribution to state propaganda, would reflect on their past indiscretions and understand how, and why, their life turned out the way that it did as a basis of their attitudes, previous decisions, and interpretations thereof. If the English Hypothesis provides the correct explanation of R_1 and R_2, then we would not expect to recognise any similarities between these situations, and the situation of the Japanese artist. In contrast, if the Self-Deception Hypothesis were the correct explanation of those events, then we would expect to observe pertinent analogies between how Stevens and the artist (and, let's continue to suppose, you) reflect on their pasts. Those familiar with Ishiguro's work should know where I'm going. The first-person narrator in *An Artist of the Floating World* exhibits just these kinds of characteristics, thereby confirming H' at the expense of H. The structure of the example matches the structure of (Schupbach's (2018) account of) robustness analysis exactly: one explores, via a work of fiction (or model), which of some set of hypotheses that compete to explain some past observations, is robust across a novel scenario, and the one that is so is thereby confirmed by this observation.

Examples can be multiplied. Consider the way in which (radical) feminist (H), (radical) class-based (H'), and (moderately) intersectional approaches (H^*) explain power dynamics as realised in romantic relationships. According to (a caricature of) the former two, those dynamics will track gender and class respectively. According to the latter, it's the interaction between these dimensions that we should focus on, if we want to explain such dynamics. Suppose that, in one's relatively limited relationship experience (E), these hypotheses compete with one another, but the actual experiences don't discriminate between them. One way to tell between them would be to explore a world of literary fiction where they do come apart. For example, one could explore a work that focuses on multiple stages of a relationship between a relatively well-off woman and a relatively less well-off man, starting from their high school experiences in Sligo and continuing during their university experience in Dublin. If the intersectional hypothesis is correct, then we would expect to see a shift in the power dynamics as the relationship develops (R_n), with each dimension of privilege interacting with the other (in an albeit complicated manner), in part as a result of how they are embedded within the different wider social contexts. If either of the other hypotheses is correct, we would not expect to see such a shift ($\neg R_n$) (recall that I am caricaturing them, at least to some extent). This is exactly what we see in Rooney's *Normal People*, which thus provides the test case which thereby discriminates between the competing hypotheses, confirming H^* at the expense of its competitors.

Or consider Haldeman's *Forever War*. Suppose that one wants to explain why R_1: a Vietnam war veteran, upon returning to the America, reported: 'I had the distinct feeling (common among returning veterans I think) that this was not my country, not my time' (Heinemann, quoted in Isaacs (1997, 12), quoted in Wiltsche (2021, 3691)). Could this experience be explained by factors specific to the Vietnam war, the way it was reported in the media at the time, the public attitude to the war, the politicians' reaction to these factors, and so on (H)? Or is this experience to be explained by the more general disconnect that occurs between soldiers and the civilians they leave behind, where this disconnect is understood in terms of the distance between the metaphorical "life lines" of the two: the soldiers' forking when they enter the service, continuing to diverge when they are on the battlefield, and thus failing to align when they are reunited (H')? Again, both serve as potential explanations of the veterans' experience on returning home from the actual war, so how we can discriminate between them? Well, we can turn to a work of literary fiction which, by construction, would tell between the two. And this is exactly what happens in *Forever War*, where the second explanatory hypothesis is realised by Haldeman's use of relativistic time dilation (thereby constructing an extreme divergence between the soldiers' and civilians' "life lines"), which in turn is used to explore the sense of alienation that is felt by veterans on their return. Again: by exploring the world of fiction we can discriminate between explanatory hypotheses (which compete with respect to our actual experiences), and confirm one at the expense of the other.[23]

Once we start to apply this structure to our reading of works of literary fiction – exploring the idea that they can be used to discriminate between competing explanatory hypotheses – it is surprisingly easy to find examples that fit the mould. Are you interested in how British/American academics are able to adapt to their respective academic contexts despite the significant cultural differences? Then observe how your potential explanations work in the world of Lodge's *Changing Places*. This, I think, provides good reason to believe that this is (one of) the mechanisms through which we learn from works of literary fiction. And when we learn in this way, by confirming one of our potential explanations for our actual experiences, we thereby gain knowledge about the actual world.

8.4 Objections and Replies

In support of aesthetic cognitivism I have presented, via several examples in each case, two different analogies between how we learn from fiction and aspects of scientific practice. Recall that the cognitivist is committed to two claims: that we learn from art (which I interpret as gaining propositional knowledge about the actual world); and that this cognitive value contributes to its aesthetic value. As applied to my previous discussion, an anti-cognitivist would need to deny either that what I have described is properly understood as a cognitive good in this sense, or that it plays an aesthetic role. In this section, I address these concerns in reverse order, and in the next I turn to broader reflections on what lessons, concerning the relationship between art and science more generally, we should draw from my discussion.

Taking the aesthetic claim first, the anti-cognitivist might grant, for the sake of argument, that we learn from works of literary fiction in a manner like the one I sketched above, but argue that this learning is irrelevant to the aesthetic value of the works in question. As discussed in Section 8.2.1, I think that this strategy loses its teeth once the cognitive value in question is characterised as propositional knowledge about the actual world, and as such, it's not clear to me how viable this objection is. However, it is still worth addressing it by reflecting on the details of the examples I have offered. The way in which I (following Elgin 2017) have argued that we learn from *Pride and Prejudice* concerns exploring the interplay between social and personal motivations and how they impact our understanding of others within the microcosm of the world of the novel; the way in which I have argued we learn from *Remains of the Day* concerns how we understand how Stevens', perhaps misguided, sense of dignity has impacted events in his life, and how he interpreted (and interprets) them. I further claimed: that we learn by comparing the similarities between *Remains of the Day* with *An Artist of Floating World*; and that we learn about intersectionality by reading *Normal People*, and the alienation of veterans by reading *Forever War* (and I suggested that we learn about academic cultures from *Changing Places*). Each of these lessons concern aspects of the novels in question that are central to their aesthetic value. These are not, in any sense, "accidental" to the works; they are their central themes, and they are not, at least in any obvious sense, "external to their status as art works". *Pride and Prejudice* is a novel of manners, Ishiguro explores the unreliable narrator, the power swings between Marianne and Connell are at the heart of *Normal People*, and who would deny that part the value of *Forever War* lies in how it explores William Mandella's sense of alienation when he returns from war? So whilst I do not take myself to have conclusively established that the fact that we learn from these works in the manner described previously is, in itself, an aesthetic virtue, I take it that the onus is on the anti-cognitivist to demonstrate that it is not.

Moving onto how the anti-cognitivist might object to my epistemic claim; that we can learn from works of literary fiction. There are at least two ways that one might raise for my treatment of the above examples.[24] The first concerns the *precision* of the knowledge that we gain from these works and the second concerns whether we learn from the works *alone*. I address both in turn.

I have argued that from *Pride and Prejudice* we learn how particular social and facts (relating to social expectations, emotions, self-understanding, and so on) can contribute to (mis)interpreting the motivations of others in romantic relationships. Such a lesson is relatively vague. In fact, it is reminiscent of Stolnitz's lament, when searching for a lesson about the actual world that we learn from the novel, that: 'Stubborn pride and ignorant prejudice keep attractive people apart [...] Can this be all there is? From one of the world's great novels?' (1992, 194). Stolnitz takes this as a *reductio* of aesthetic cognitivism: if this is all we learn from *Pride and Prejudice* then we would be better off looking for its aesthetic value elsewhere.

But this is misguided for two reasons. First, whilst this might be the *general lesson* of the work, through seeing how it is realised in the fictional world associated with the fiction we can also come to understand how it is realised (albeit in a more complicated way) in actual systems in the world. From our knowledge gained by reading the work we can do more than come to understand that pride and prejudice can undermine romantic relationships in general, we can also understand the mechanisms (how we are led to misunderstand the motivations of others, and so on) through which this works.[25] And whilst these mechanisms may be embedded within a more complex nexus in any actual system, this doesn't mean that they aren't present at all.[26]

Second, even if this were all that is to be learnt from Austen's novel, in what sense should it be understood as banal, or trivial? Well, we might contrast it with the cognitive goods that science provides: one might think that the vagueness associated with Stolnitz' lesson stands in sharp contrast to the precise knowledge that we gain through our scientific endeavours. And, if the lessons that we learn from literary fiction are inherently vague in comparison, then this might be reason to deny that they should be properly deemed of cognitive value. I think this rests on a misunderstanding of scientific practice itself. Consider again, the "lessons" that we learn from our experimenting, physically and in thought, on pendula in order to draw lessons about acceleration due to gravity on the earth's surface, and the relationship between their length and period. Whilst we may gain precise knowledge about the tightly controlled experimental setups, when we attempt to generalise from them to systems "in the wild", we inevitably introduce a degree of imprecision. We cannot infer that the value of g in general is *exactly* the same as the value measured at a particular location (not least because there is no value in general, but also even in our physical experiment we have to accommodate measurement error due to the impact of other forces), and so some level of imprecision must be introduced in our experiment-to-world reasoning. The same applies to reasoning from model systems, as has been discussed extensively by example (Nguyen 2020; Nguyen and Frigg 2020). It is a mistake to think that models provide us with precise information about the world (e.g. it is a mistake to think that they are isomorphic to their target systems), given the fact that the former include all kinds of idealisation assumptions (Frigg and Nguyen 2020, Chapter 4). But once we understand that they are to be interpreted rather less precisely, we can recognise their cognitive value is imprecise, but knowledge nevertheless (Frigg and Nguyen 2019). I see no reason why the same fails to apply to the lessons drawn from the works of literary fiction.

The second objection the anti-cognitivist can make to my treatment of the epistemic claim concerns the extent to which the works of literary fiction *alone* deliver the cognitive goods I have described. One way of putting this objection would be to consider some alternative works of fiction to the ones I described in Section 8.3.2 (a similar objection and response can be generated for the discussion of the isolation mechanism). It is not difficult to imagine works of fiction that: confirm the hypothesis that it is precisely his Englishness that explains why Stevens is the way he is; that confirm the idea that gendered power imbalances trump class-based ones; or that confirm the claim that there were factors specific to the Vietnam war that explain the sense of alienation experienced by homecoming troops. And since these works would confirm the opposite hypotheses to the ones presented as confirmed in my above discussion, doesn't this rather pull the rug from under my claim? I think that this rests of a mistaken idea of what it takes to confirm a hypothesis, and again this can be illustrated by comparison with the examples from scientific practice in Section 8.2.2.2. With some ingenuity, one *could* construct a measurement scenario according to which an inorganic material fails to exhibit Brownian motion (experiments can be mistaken after all), or one *could* construct a model according to which it's precisely the no agent-level interactions that explain why Volterra's model exhibits the Volterra property (imagine an agent-level model gerrymandered to not exhibit it). This suggests that we should understand the cognitive value of scientific individual practices (like physical and thought experiments, and experimental and model-based robustness analysis) as *contributing* to our knowledge, against a background of theoretical assumptions and commitments, rather than directly providing such value *all by themselves*. Indeed there is nothing "built into" a model that will guarantee that it is successful: our confidence in the result of a modelling endeavour, typically at least, requires reference to the research programme in which it is embedded (cf. Frigg and Nguyen 2022).[27] And again, I see no reason why the same fails to apply to the lessons drawn from the works of literary fiction above.

8.5 Conclusion

The final discussions in the previous section relied on a particular conception of scientific practice, according to which science delivers cognitive goods that are less precise, and rely more on wider theoretical considerations, than one might initially think. This provides me with a segue into the final point I want to make in this chapter.

A defender of the arts may object to my argument along the following lines: any attempt to account for the value of works of literary fiction in the manner sketched above, in terms of the scientific experiment and robustness analysis, smacks of scientistic imperialism (they might point to my use of roman variables, "*H*"s denoting "hypotheses" and "*R*"s denoting "explananda", used to my own chagrin, as indicative of such an approach). But the appropriate lesson to draw from the above discussion is not that the cognitive value of art should be assimilated into the scientific regime. The claim is not that "science" has a monopoly on cognitive value, and therefore cognitivists should attempt to demonstrate how works of art can be forced into the structures that philosophers of science extract during their rational reconstructions of scientific practice. Rather, I think we should recognise both artistic and scientific practices as two aspects of a more general humanistic way of dealing with the complexity of the world via simplified constructions. Of course whilst there may be important underlying differences between these practices, and the fact that experimentalists and artists alike construct shielded systems in order to explore how certain features of the world operate when isolated from others (as Austen herself notes, for the sorts of features she wants to explore, 'three or four families in a country village is the very thing to work on' (quoted in Elgin 2017, 237)) doesn't mean that artists are doing science any more than it means that scientists are creating art, and neither does exploring how certain explanatory ideas work when imagined to hold in fictional scenarios.[28] But it does mean that they are both engaged in common projects, describable, at a certain level of abstraction, as sharing a common structure. And if we, as philosophers of science and art, can recognise this structure (which includes an epistemologically modest way of recognising how science works, as imprecise and holistic), then we can come to a better understanding of the epistemological position we find ourselves in.

To conclude then, I provided a defence of aesthetic cognitivism as applied to literary fiction by demonstrating structural similarities between how we can learn from them on the one hand and scientific experiments and robustness analysis on the other. This was demonstrated via the use of examples. By applying such frameworks to those cases, I hope to have furthered our understanding of their value and function, and indeed our appreciation of them as works of literature. Moreover, I hope that this can be generalised to other works of literary fiction, and art more generally, where we want to explicate the mechanism through which we learn from them (if we, in fact, do so). My wider aim is to have demonstrated the viability of taking lessons from the philosophy of science across the border to aesthetics and the philosophy of art more generally. If, as I alluded to above, science and art are better seen as part of a wider humanistic project (at least at some appropriate level of abstraction), then we can provide a rich set of epistemological tools for understanding the cognitive value of art, and how it fits into a unified epistemological framework.

Notes

1 james.nguyen@philosophy.su.se. Department of Philosophy, Stockholm University, Sweden. Institute of Philosophy, School of Advanced Study, University of London, UK; and Centre for Philosophy of Natural and Social Science, London School of Economics and Political Science, UK. I am grateful to the Jeffrey Rubinoff Sculpture Park and the Jacobsen Fund at the University of London for their support while writing this chapter. I have received very useful feedback from audiences at: the Institute of Philosophy, University of London; the Institute of Philosophy, University of Bern; and the First Workshop on the Aesthetics of Scientific Experiments at the University of Cambridge. Thanks also to Margherita Harris for stimulating discussion of the material, and Milena Ivanova and Alice Murphy for useful comments, and for inviting me to participate in this wonderful project.

2 I make no claims of novelty in this regard. This point has been made forcefully by Elgin (2017, chapters 11 and 12), which I extend here. For a critical discussion of this approach see (Egan 2016). Murphy (2020) furthers the debate by emphasising the literary qualities of scientific thought experiments.

3 See also (Reiss 2015).

4 I use the term "actual world" broadly here, to include both the natural and social (including moral). I also include what we might call "the world of thought", although I don't explicate this in terms of describing our *concepts*, but rather our *thought processes*. For more on the idea that we can learn about our concepts from works of literary fiction see (Nussbaum 1992; John 1998).

5 Depending on one's preferred account of fiction we could make sense of this in terms of what is true-in-the-possible-world described by the fiction, or what is true-in-the-game-of-make-believe associated with the fiction. Either way, what is true-in-the-fiction goes beyond what is explicitly stated in the work.

6 For influential anti-cognitivist arguments see (Stolnitz 1992; Lamarque and Olsen 1994; Lamarque 2006). It is also worth mentioning that the anti-cognitivist claim can be further motivated by general idea, often associated with Kant, that aesthetic value is in some sense "autonomous" (see, e.g. Haskins 1989, for a discussion). I cannot address this general issue here.

7 For useful discussions distinguishing these cognitive goods as applied to the cognitive value of art see (Gaut 2005, 2007; Gibson 2008; Peels 2019).

8 Although by this, as will become clear, I don't mean that we learn from art simply by increasing our stock of justified, true, beliefs (let alone beliefs whose contents can be precisely stated). I mean something more subtle, something that allows for learning by increasing our confidence in various claims, for learning about how our claims fit together, and, in particular, for learning how certain features of the world can play (partial) explanatory roles in actual systems.

9 An alternative approach is to grant the association of such cognitive goods with works of art, but to argue that they are in some sense 'accidental', or fail to be 'internal to their status as artworks' (Gibson 2008, 4). But learning about the fictional world associated with the text involves clearly seems to involve learning something non-accidental or internal to the work's status as art: we wouldn't want to deny that in reading *Changing Places* we learn that "Rummidge is a university in England in the novel *Changing Places*", or

that we can (at least to some extent) gain experiential knowledge associated with what it *would be like* to be in a, e.g., the narrator's position in a fictional world. Nor would we claim that such learning is accidental or external to the work's status as a literary fiction. As a result, I take it that the anti-cognitivist would be better served to argue that such learning is aesthetically irrelevant, rather than claiming it to be accidental and/or external to the work, or denying its existence in the first place.

10 The question of whether you have successfully measured the quantify of interest in an experiment concerns the experiment's 'internal validity' a term originally introduced in the social sciences in (Campbell 1957).

11 For my current purposes, I'm not distinguishing between thought experiments and models, compare, e.g., (Murphy 2022; Stuart 2022).

12 For more discussion on the analogy between thought experiments (particularly simulations) and physical experiments see (Guala 2002; Winsberg 2003; Morgan 2005; Parker 2009; Arcangeli 2018; French and Murphy 2021).

13 The inference from physical experiment to a wider set of systems is discussed in terms of the experiment's 'external validity' (Campbell 1957). My thinking about the inference from experiment to systems 'in the wild' is greatly influenced by (Cartwright 1983, 1999). For more on the inference from models and thought experiments to such systems see (Frigg and Nguyen 2020) (and of particular relevance in the current context (Nguyen 2020) and (Salis and Frigg 2016; Stuart et al. 2017), respectively).

14 At least insofar as I endorse the structure of how he reconstructs the reasoning. I don't think my discussion here relies on the details, just that this style of inference provides us with knowledge about the actual world, a claim that most involved in the debate agree with, even if they disagree with respect to the details (for useful discussions, see Levins 1966; Orzack and Sober 1993; Weisberg 2006; Kuorikoski et al. 2010; Odenbaugh and Alexandrova 2011; Parker 2011; Hudson 2013; Harris 2021). Of course, if it turns out that robustness analysis does not deliver such a cognitive good in the scientific context, then my argument that it does in the artistic context will be blocked. But a result undermining this significant aspect of scientific practice should/would cause us to reconsider our account of cognitive goods in the first place.

15 My claim that works of literary fiction are cognitively valuable in virtue of displaying how features of the world act in isolation draws heavily from (Elgin 2017, chapters 11 and 12).

16 Although of course wealth does play an important role, it is not one that is necessarily motivated by the need to earn enough to live, and Elizabeth doesn't worry about whether Mr Darcy will need to move across the Atlantic for work.

17 For example, we are provided with relatively little information about Miss Kenton's mental life (beyond that, we can infer from Stevens' description of his perspective); next to no information about the social situation beyond Darlington Hall and Stevens' road trip; and even with respect to Steven' mental life we don't explore what he thinks about religion, art, or culture, except insofar as these impact the way in which his self-perception shapes how he interprets and fails to embrace the possibilities he faces.

18 There is another important cognitive good to be found in *Remains of the Day*, namely we learn how to interpret the actions and behaviour of others. In being encouraged to infer, from what we are told from Stevens' first-person perspective, to how he has misinterpreted certain events, and how he (in a sense) lies to himself in order to support his self-perception, we thereby develop our skills in applying this inferential process to actual people. I will not explore this cognitive good here, since I think it is better characterised as a cognitive skill rather than an information-based account of cognitive value. But this provides evidence for a pluralist approach to the cognitive value of art.

19 And as applied to the sorts of lessons I'm focusing here, see (Robinson 1995), although she does not talk about the mechanism that generates them in the way that I do.
20 I return to this example because I believe in a pluralistic aesthetic cognitivism: works of art, even the very same work, can provide multiple cognitive goods, even via multiple different mechanisms.
21 The ambiguity of whether Stevens himself realises this is part of what makes *Remains of the Day* a wonderful novel. My argument here does not rely on interpreting the character either way.
22 I take it that most readers will understand what I mean by this, and some may object to the nationalistic stereotyping. As should become clear, I am not endorsing the idea that this concept is legitimate, at least in any rich sense. All I require is that it plays a *potentially* explanatory role.
23 I take this example from Wiltsche (2021), who also argues that it holds cognitive value. However, he argues that this value is realised in the way that the work allows the reader to come 'to understand what it would mean for human beings to inhabit a world in which relativistic effects like time dilation directly interferes with their everyday lives' (2021, 3691). I think what Wiltsche says is true (recall that I am a pluralist about the cognitive value of literary fiction), but I don't think this exhausts the cognitive value of the work. In particular, I think it helps explain why veterans feel alienated on their return, in the manner sketched above.
24 For a related general discussion about learning from scientific fictions see (Frigg and Nguyen 2021).
25 In this sense, the propositional cognitive value that we gain from works of art may be intimately connected to the experiential or emotional value (or skills or understanding) that others have proposed as the sources of art's cognitive value. But in which case I stress that this should apply equally to the sorts of scientific contexts I have been talking about, and not taken to differentiate between art and science (with the latter being mistakenly thought to provide propositional knowledge in some straightforward sense). I am grateful to Alice Murphy for encouraging me to be explicit about this.
26 For a discussion of this in the scientific context see (Potochnik 2017, Chapter 2) and (Nguyen 2020).
27 I say "typically at least" here because some (but not all) thought experiments might be best thought of as delivering cognitive value "in isolation" so to speak; in particular thought experiments that test the (in)consistency of various propositions. See (Skaf 2021) for a discussion.
28 Or turning this point on its head: it does, but these claims hold both ways.

Bibliography

Arcangeli, M. (2018) 'The Hidden Links Between Real, Thought and Numerical Experiments', *Croatian Journal of Philosophy*, 18(1), 3–22.

Budd, M. (1993) 'How Pictures Look. Book Section', in D. Knowles and J. Skorupski (eds.) *Virtue and Taste*. Oxford: Blackwell, pp. 154–175.

Bueno, O., Darby, G., French, S. and Rickles, D. (2017) *Thinking about Science, Reflecting on Art: Bringing Aesthetics and Philosophy of Science Together*. Oxford and New York, NY: Taylor & Francis. https://books.google.se/books?id=hxIwDwAAQBAJ.

Campbell, D. T. (1957) 'Factors Relevant to the Validity of Experiments in Social Settings', *Psychological Bulletin*, 54(4), 297–312.

Cartwright, N. (1983) *How the Laws of Physics Lie*. Oxford: Oxford University Press.

Cartwright, N. (1999) *The Dappled World: A Study of the Boundaries of Science*. Cambridge: Cambridge University Press.

Egan, D. (2016) 'Literature and Thought Experiments', *The Journal of Aesthetics and Art Criticism*, 74(2), 139–150. https://doi.org/10.1111/jaac.12270.

Elgin, C. Z. (2017) *True Enough*. Cambridge, MA, and London: MIT Press.

French, S. (2003) 'A Model-Theoretic Account of Representation (or, I Don't Know Much About Art ... But I Know It Involves Isomorphism)', *Philosophy of Science*, 70, 1472–1483.

French, S. and Murphy, A. (2021) 'The Value of Surprise in Science', *Erkenntnis*, 1–20. https://doi.org/10.1007/s10670-021-00410-z.

Frigg, R. and Nguyen, J. (2016) 'The Fiction View of Models Reloaded', *The Monist*, 99(3), 225–242. https://doi.org/10.1093/monist/onw002.

Frigg, R. and Nguyen, J. (2019) 'Mirrors without Warnings', *Synthese*, 198(3), 2427–2447. https://doi.org/10.1007/s11229-019-02222-9.

Frigg, R. and Nguyen, J. (2020) *Modelling Nature: An Opinionated Introduction to Scientific Representation*. Cham: Springer.

Frigg, R. and Nguyen, J. (2021) 'Seven Myths about the Fiction View of Models', in A. Cassini and J. Redmond (eds.) *Models and Idealizations in Science: Artifactual and Fictional Approaches*. Cham: Springer Verlag, pp. 133–157.

Frigg, R. and Nguyen, J. (2022) 'DEKI and the Mislocation of Justification: A Response to Millson and Risjord', in I. Lawler, K. Khalifa and E. Shech (eds.) *Scientific Understanding and Representation: Modeling in the Physical Sciences*. New York, NY, and London: Routledge.

Gaut, B. (2005) 'Art and Knowledge', *The Oxford Handbook of Aesthetics*. Oxford: Oxford University Press. https://doi.org/10.1093/oxfordhb/9780199279456.003.0025.

Gaut, B. (2007) *Art, Emotion and Ethics*. Oxford: Oxford University Press.

Gibson, J. (2008) 'Cognitivism and the Arts', *Philosophy Compass*, 3(4), 573–589. https://doi.org/10.1111/j.1747-9991.2008.00144.x.

Goodman, N. (1976) *Languages of Art*. 2nd ed. Indianapolis, IN; Cambridge: Hacket.

Guala, F. (2002) 'Models, Simulations, and Experiments', in L. Magnani and N. J. Nersessian (eds.) *Model-Based Reasoning: Science, Technology, Values*. Boston, MA: Springer US, pp. 59–74. https://doi.org/10.1007/978-1-4615-0605-8_4.

Haskins, C. (1989) 'Kant and the Autonomy of Art', *The Journal of Aesthetics and Art Criticism*, 47(1), 43–54. http://www.jstor.org/stable/431992.

Hudson, R. (2013) *Seeing Things: The Philosophy of Reliable Observation*. Oxford; New York, NY: Oxford University Press.

Isaacs, A. R. (1997) *Vietnam Shadows: The War, Its Ghosts, and Its Legacy*. American Moment. Baltimore, MD, and London: Johns Hopkins University Press. https://books.google.se/books?id=RH5uAAAAMAAJ.

Ivanova, M. (2017) 'Aesthetic Values in Science', *Philosophy Compass*, 12(10), e12433. https://doi.org/10.1111/phc3.12433.

Ivanova, M. and French, S. (2020) *The Aesthetics of Science: Beauty, Imagination and Understanding*. Routledge Studies in the Philosophy of Science. New York, NY: Taylor & Francis. https://books.google.se/books?id=LszKDwAAQBAJ.

John, E. (1998) 'Reading Fiction and Conceptual Knowledge: Philosophical Thought in Literary Context', *Journal of Aesthetics and Art Criticism*, 56(4), 331–348. https://doi.org/10.2307/432124.

Kuorikoski, J, Lehtinen, A and Marchionni, C. (2010) 'Economic Modelling as Robustness Analysis', *The British Journal for the Philosophy of Science*, 61(3), 541–567. https://doi.org/10.1093/bjps/axp049.

Lamarque, P. (2006) 'Cognitive Values in the Arts: Marking the Boundaries', in M. Kieran (ed.) *Contemporary Debates in Aesthetics and the Philosophy of Art*. Blackwell, pp. 127–139. https://www.wiley.com/en-kr/Contemporary+Debates+in+Aesthetics+and+the+Philosophy+of+Art-p-9781405102391

Lamarque, P. and Olsen, S. H. (1994) *Truth, Fiction, and Literature: A Philosophical Perspective.* Oxford: Oxford University Press.

Levins, R. (1966) 'The Strategy of Model Building in Population Biology', *American Scientist,* 54(4), 421–431. http://www.jstor.org/stable/27836590.

Lewis, D. (1983) 'Truth in Fiction', *Philosophical Papers Volume I.* Oxford: Oxford University Press. https://doi.org/10.1093/0195032047.003.0015.

Morgan, M. S. (2005) 'Experiments versus Models: New Phenomena, Inference and Surprise', *Journal of Economic Methodology,* 12(2), 317–329. https://doi.org/10.1080/13501780500086313.

Murphy, A. (2020) 'The Aesthetic and Literary Qualities of Scientific Thought Experiments', in M. Ivanova and S. French (eds.) *The Aesthetics of Science: Beauty, Imagination and Understanding*, pp. 146–166. https://www.taylorfrancis.com/chapters/edit/10.4324/9780429030284-8/aesthetic-literary-qualities-scientific-thought-experiments-alice-murphy

Murphy, A. (2022) 'Imagination in Science', *Philosophy Compass,* 17(6), e12836. https://doi.org/10.1111/phc3.12836.

Nguyen, J. (2020) 'It's Not a Game: Accurate Representation with Toy Models', *The British Journal for the Philosophy of Science,* 71(3), 1013–1041. https://doi.org/10.1093/bjps/axz010.

Nguyen, J. and Frigg, R. (2020) 'Unlocking Limits', *Argumenta*, 6(1), 31–45. http://philsci-archive.pitt.edu/20920/.

Nussbaum, M. C. (1992) *Love's Knowledge: Essays on Philosophy and Literature.* Oxford Paperbacks. USA: Oxford University Press. https://books.google.se/books?id=CXQ8DwAAQBAJ.

O'Loughlin, I. and McCallum, K. (2019) 'The Aesthetics of Theory Selection and the Logics of Art', *Philosophy of Science,* 86(2), 325–343. https://doi.org/10.1086/701953.

Odenbaugh, J. and Alexandrova, A. (2011) 'Buyer Beware: Robustness Analyses in Economics and Biology', *Biology and Philosophy,* 26(5), 757–771. https://doi.org/10.1007/s10539-011-9278-y.

Orzack, S. H. and Sober, E. (1993) 'A Critical Assessment of Levins's the Strategy of Model Building in Population Biology (1966)', *The Quarterly Review of Biology,* 68(4), 533–546. http://www.jstor.org/stable/3037250.

Parker, W. S. (2009) 'Does Matter Really Matter? Computer Simulations, Experiments, and Materiality', *Synthese,* 169(3), 483–496. http://www.jstor.org/stable/40271305.

Parker, W. S. (2011) 'When Climate Models Agree: The Significance of Robust Model Predictions*', *Philosophy of Science,* 78(4), 579–600. http://www.jstor.org/stable/10.1086/661566.

Peels, R. (2019) 'How Literature Delivers Knowledge and Understanding, Illustrated by Hardy's Tess of the D'Urbervilles and Wharton's Summer', *The British Journal of Aesthetics,* 60(2), 199–222. https://doi.org/10.1093/aesthj/ayz040.

Potochnik, A. (2017) *Idealization and the Aims of Science.* Chicago, IL: University of Chicago Press.

Reiss, J. (2015) 'A Pragmatist Theory of Evidence', *Philosophy of Science,* 82(3), 341–362. http://www.jstor.org/stable/10.1086/681643.

Robinson, J. (1995) 'L'éducation Sentimentale', *Australasian Journal of Philosophy,* 73(2), 212–226. https://doi.org/10.1080/00048409512346551.

Salis, F. and Frigg, R. (2016) 'Capturing the Scientific Imagination', in P. Godfrey-Smith and A. Levy (eds.) *The Scientific Imagination.* New York, NY: Oxford University Press.

Schupbach, J. N. (2018) 'Robustness Analysis as Explanatory Reasoning', *The British Journal for the Philosophy of Science,* 69(1), 275–300. https://doi.org/10.1093/bjps/axw008.

Skaf, R. El. (2021) 'Probing Theoretical Statements with Thought Experiments', *Synthese,* 199(3–4), 6119–6147. https://doi.org/10.1007/s11229-021-03063-1.

Stolnitz, J. (1992) 'On the Cognitive Triviality of Art', *British Journal of Aesthetics,* 32(3), 191–200. https://doi.org/10.1093/bjaesthetics/32.3.191.

Stuart, M. T. (2022) 'Sharpening the Tools of Imagination', *Synthese,* 200(6), 1–22. https://doi.org/10.1007/s11229-022-03939-w

Stuart, M. T., Fehige, Y. and Brown, J. R. (2017) *The Routledge Companion to Thought Experiments.* Routledge Philosophy Companions. Oxford and New York, NY: Taylor & Francis. https://books.google.se/books?id=BMItDwAAQBAJ.

Toon, A. (2010) 'Models as Make-Believe', in R. Frigg and M. Hunter (eds.) *Beyond Mimesis and Convention: Representation in Art and Science.* Berlin: Springer, pp. 71–96.

Volterra, V. (1926) 'Fluctuations in the Abundance of a Species Considered Mathematically', *Nature,* 118, 558–560.

Walton, K. L. (1990) *Mimesis as Make-Believe: On the Foundations of the Representational Arts.* Cambridge MA: Harvard University Press.

Weisberg, M. (2006) 'Robustness Analysis', *Philosophy of Science,* 73(5), 730–742. http://www.jstor.org/stable/10.1086/518628.

Weisberg, M. and Reisman, K. (2008) 'The Robust Volterra Principle', *Philosophy of Science,* 75(1), 106–131.

Wiltsche, H. A. (2021) 'The Forever War: Understanding, Science Fiction, and Thought Experiments', *Synthese,* 198(4), 3675–3698. https://doi.org/10.1007/s11229-019-02306-6.

Winsberg, E. (2003) 'Simulated Experiments: Methodology for a Virtual World', *Philosophy of Science,* 70(1), 105–125. https://doi.org/10.1086/367872.

9 Something from Nothing

"Non-discovery" and Transformations in High Energy Experimental Physics at the Large Hadron Collider

Sophie Ritson

9.1 Introduction

After a discovery claim is established, the world is reported to be different; by contrast, when something is lost that was never had, seemingly nothing has changed. Drawing on interviews with experimental physicists involved in the search for, and (thus far) non-discovery of, supersymmetry (SUSY) at the Large Hadron Collider (LHC), this chapter will explore transformations in epistemic strategies. In particular, the chapter explores how the absence of direction in research strategies, following the negative results at the LHC, is transforming the epistemic strategies of experimental particle physicists in the ATLAS and CMS experiments at the LHC, in an effort to generate ugly experimental results.

Prior to the start of data collection, many in the high energy physics community strongly expected that evidence for SUSY would be observed at the LHC. To date, these expectations have not been realised. I will show that whilst the negative results have not transformed the ontology of particle physics, there has been a transformation in epistemic strategies away from targeted searches for evidence of SUSY (and other beyond standard model (BSM) physics) to attempts to find evidence for ugly results and 'unconceived alternatives' (Stanford 2006). Epistemic strategies that aim at "unconceived alternatives" are apparent in the shift towards "model independent searches" and standard model measurements. This chapter will focus on one such example: the attempts to measure the self-coupling of the Higgs boson. These attempts are in part motivated by the possibility that the measured result will disconfirm the value predicted by the standard model, thereby providing a path to unconceived and alternative new physics.

Stanford (2019) has argued that, due to conservative attitudes, today's scientific communities are 'less effective' than their predecessors 'in developing fundamentally novel theoretical conceptions of nature' (p. 3931). The current situation in particle physics provides an opportunity to examine the creativity of the highly collaborative experimental particle physics community, where the transformation in epistemic

DOI: 10.4324/9781003263920-10

strategies indicates direct attempts to find evidence for "unconceived alternatives", or for disconfirming and "ugly" experimental results, that could provide fundamentally novel concepts.

This chapter will proceed as follows; it will first give an outline of two of the experiments at the LHC, the ATLAS, and CMS experiments, including an overview of the attempts to find evidence for SUSY and the (to date) lack of evidence that has been found. The chapter will then detail attempts to measure the self-coupling of the Higgs boson in the ATLAS and CMS experiments and show that, in this very specific context, ugliness can have epistemic value: its nature is transformative.

9.2 ATLAS and CMS Experiments and the Search for SUSY

The LHC at CERN is one of the largest and most complex experiments ever built, consisting of a 27-km ring in which protons are accelerated and made to collide in bunches of proton-proton collisions in four detectors. Each of these detectors was independently built and is run by a large experimental collaboration: the ALICE, ATLAS, CMS, and LHCb experimental collaborations. The ATLAS and CMS experimental collaborations built multipurpose detectors, both designed with the same aims: to understand the origin of electroweak symmetry breaking, to search for BSM physics, and to perform precision measurements of processes within and beyond the standard model (CMS Collaboration 2002; ATLAS Collaboration 2003). At the time of writing, each of the ATLAS and CMS collaborations has over 3000 members spread across the world at different research institutions.

It is impossible for the detectors to "observe" directly, or interact with, the short-lived particles, such as Higgs bosons, created by the collisions. Instead, the multipurpose detectors are optimised for the decay products. For example, the initial evidence found for the Higgs boson was a measurement of an excess of two of its possible decays: photons pairs and Z boson pairs. The Z boson is also not directly observable; instead, it is inferred from the decay products (electron pairs or muon pairs). The decay products are also not directly observable, though they do interact with the detector to produce electronic signals. Furthermore, the decay products relevant to the search or measurement performed must be inferred from the satisfaction of specific data selection criteria.[1]

The search for physics beyond the standard model is driven by the perceived incompleteness of the standard model, as well as claims of conceptual problems in the standard model. The standard model is an assemblage of theoretical and experimental insights describing particles and their interactions that was developed by experimental and theoretical physicists beginning in the 1970s. The contemporary standard model includes three fundamental forces (or interactions) – the strong, weak, and electromagnetic forces – and is comprised of two gauge theories; namely, the electroweak theory which unifies and describes the weak and electromagnetic interactions, and the theory of quantum chromodynamics (QCD) which describes strong interactions. The standard model has been extraordinarily successful in that it has been confirmed by thousands of experiments, to very high degree of precision. However, there is a long history of claims that the standard model is incomplete as it does not contain gravity or dark matter and cannot be used to answer questions such as: what determines the value of the particle masses? What is the nature of astrophysical dark matter? And what is the quantum theory of gravity?

There also exist a large number of candidate approaches to BSM physics that introduce new concepts, particles, or interactions. One approach, which has resulted in a large number of theories, is supersymmetry, or "SUSY" for short. Broadly, SUSY theories predict a partner particle for each particle in the standard model. The "standard" arguments for SUSY, that I will reproduce here, are that SUSY offers an "explanation" for the mass of the Higgs boson, or, more specifically, how light it is. A light Higgs boson is considered improbable as the interactions between the Higgs and other standard model particles would result in a heavy Higgs boson. The extra particles predicted by some SUSY approaches would cancel out the contributions to the Higgs boson's mass from their standard model partners, making a light Higgs boson possible.

SUSY also introduces further symmetry to the standard model. Fermions all have half of a unit of spin, whilst the bosons have 0, 1, or 2 units of spin. SUSY predicts that each of the particles in the standard model has a partner with a spin that differs by half of a unit. Bosons are accompanied by fermions and vice versa. Linked to their differences in spin are differences in their collective properties. Several SUSY theories predict the lightest supersymmetric particle to be stable and electrically neutral and to interact weakly with the particles of the standard model, thereby predicting experimentally detectable dark matter candidates. If supersymmetric particles were included in the standard model, the interactions of its three forces – electromagnetism and the strong and weak nuclear forces – could have the same strength at very high energies, which is considered evidence for a unified theory. These arguments were very convincing to many in the theoretical and experimental physics communities. In 2014, experimental high energy physicists Joseph Lykken, a theoretical physicist at the Fermi National Accelerator Laboratory, and Maria Spiropulu, a particle physicist at the California Institute of Technology and a member of CMS, claimed that 'it is not an exaggeration to say that most of the world's particle physicists believe that supersymmetry must be true—the theory is that compelling' (Lykken and Spiropulu 2014). In 2016, Spiropulu further reflected on attitudes towards SUSY: 'we had figured it all out … If you ask people of my generation, we were almost taught that supersymmetry is there even if we haven't discovered it. We believed it' (quoted in Wolchover 2016).

Spiropulu's reflections came shortly after the second major phase of data collection by the ATLAS and CMS experiments, the so-called Run 2. No evidence for SUSY had been found in Run 1, or indeed at the LHC's predecessor, the Large Electron Positron Collider that ran at CERN from 1989 to 2000, or at the Tevatron accelerator at Fermilab which ran until 2011. In the lead-up to Run 2, astrophysicist Adam Frank and theoretical physicist Marcelo Gleiser wrote an opinion piece titled "A Crisis at the Edge of Physics" in *The New York Times*. Frank and Gleiser wrote:

> Just a few days ago, scientists restarted investigations with the Large Hadron Collider, after a two-year hiatus ... If the upgraded collider does discover supersymmetric particles, it will be an astonishing triumph of modern physics. But if nothing is found, our next steps may prove to be difficult and controversial, challenging not just how we do science but what it means to do science at all.
>
> (Frank and Gleiser 2015)

This public statement and open reflection on the epistemology of high energy physics is quite extraordinary for a number of reasons. Putting to one side the inflammatory quality of this statement, which may have been written to provoke, a connection is made from negative or null results to significant or even radical change to the epistemic practices of high energy physics. This is a claim that transformation will come from non-discovery or a discovery of nothing.

Run 2 did indeed offer no evidence for SUSY, or any candidate BSM approach. Result after result, search after search, found no evidence for BSM physics and confirmed the standard model to an increasingly higher degree of precision. Ben Allanch, a particle phenomenologist at the University of Cambridge, noted in 2018:

> none of the more exotic particles and interactions that theorists hoped to see has been forthcoming. No 'stop squarks', no 'gluinos', no 'neutralinos'. The null results are now encrusting the hull of the Standard Model, like barnacles on a beautiful old frigate, and dragging her down to the ocean floor. It looks like the centuries-long quest for top-down unification has stalled, and particle physics might have a full-blown crisis on its hands.
>
> (Allanach 2018)

There is a sense in which the ATLAS and CMS experiments were simultaneously succeeding and not succeeding. Negative or null results are an everyday matter in experimental high energy physics (and indeed in experiments in general) and there are established responses. Unlike in many other experimental sciences, null results are recognised as being worthy of publication in experimental high energy physics, and the ATLAS and CMS experiments were producing reliable, robust, and peer-reviewed negative results. However, it is difficult to say what progress was being made with respect to the second aim of the ATLAS and CMS experiments: the aim to search for evidence of physics beyond the standard model.

Part of the difficulty stems from the focus in philosophy of science and in the sciences themselves on discovery or positive claims. There is now a very large body of work in history and philosophy of science that has explored the extended processes by which discovery claims are established, and just how contingent and complex the histories of discoveries are, that has challenged claims of any simple logic of discovery.[2] As the various SUSY approaches are excluded from more and more of the phase space, more is known concerning the limits of SUSY (as boundaries are being put on where it might be reasonable to find evidence for SUSY). However, it is unclear as to what would constitute a "non-discovery" of SUSY, stemming in part from the logical asymmetry between discovery and non-discovery claims. One of the famous critiques of Popper's use of falsification as a demarcation criterion between science and pseudoscience was the extreme difficulty of falsifying existence claims. One search could provide evidence for the existence of SUSY, whereas it is extremely difficult if not impossible to claim a "non-discovery" of SUSY, not only because there are many different SUSY approaches that all make different predictions but also because it would require an unknown but huge number of experiments to explore the entire phase space. At the time of writing, the ATLAS experiment had published 167 null results from direct searches for SUSY. Such searches for SUSY put constraints on the various SUSY approaches, but there is no consensus as to how to interpret these constraints. How do you get something out of nothing? Frank and Gleiser (2015) suggested that transformation will come from a discovery of nothing; however, they did not give any hints as to the nature of the transformation.

One question that looms large is how to make future progress in experimental high energy physics. Without evidence for BSM physics, there is an absence of obvious puzzles to solve. One change, that is more incremental than outright, is a shift to creative experimentation (Ritson 2021). One area in which this is occurring is in the searches for "long-lived" particles. The ATLAS and CMS detectors were designed to find evidence for particles that decay instantly, as most BSM particles were predicted to do so. Long-lived particles interact with the detectors in ways that were not anticipated in the original design. This has required the experimental collaborations to use parts of the detector creatively, in ways in which they were never designed to be used, to search for evidence of long-lived particles. This use of the detectors has allowed the collaborations to search for evidence of the unexpected.[3] There has also been a shift to precision measurements, as opposed to direct searches or top-down approaches, an example of which I will examine in detail in the next section. What has changed is a transformation of knowledge production strategies, with an aim to use the detectors to look for evidence for previously unconceived physics scenarios in an effort to find something unpredicted or unexpected. Hopefully, as I will explore in the next section, something ugly.

9.3 Aiming for Ugliness with Measurement

During Run 1, in 2012, ATLAS and CMS both announced that they had seen evidence of a boson and measured its mass at approximately 125 GeV. This was significant evidence to suggest that ATLAS and CMS had determined the unpredicted mass of, and discovered the long-sought, last to-be-confirmed, prediction of the standard model: the Higgs boson. Confirmation of the existence of the Higgs boson is widely considered to complete the standard model. However, whilst the measurement of the mass, and subsequent measurement of the spin, of the boson and various couplings indicates that it is very likely to be the Higgs predicted by the standard model, more measurements are required to see if the Higgs does precisely play the role in the standard model that is predicted. One such measurement is the measurement of the interaction of the Higgs boson with itself or the self-coupling of the Higgs. Whilst there was no standard model prediction of the mass of the Higgs, predictions were made for the spin and various couplings.

A measurement of the Higgs trilinear self-coupling is considered achievable from an "observation" of Higgs boson pair production. The strength of the trilinear coupling influences the production rate of Higgs pairs in the standard model because trilinear self-coupling is related to the probability of the Higgs to "decay" into a pair of Higgs. The rate for this decay, and therefore the production rate, is higher if trilinear self-coupling is higher, and vice versa. The Higgs potential can be accessed experimentally through the measurement of the cross-section for Higgs pair production (where a measurement of the cross-section is a measurement of the yield of all processes that produce two Higgs bosons, including the Higgs self-coupling). The Higgs sector may be different from that which the standard model predicts, which can lead to a measurement of a different rate of di-Higgs production, yielding a different value of trilinear self-coupling. It is then possible to compare the experimentally determined self-coupling with a theoretical prediction for the standard model self-coupling.[4]

As opposed to a direct search, one aspect of the potential of this measurement is that a whole range of results is possible. What is crucial here is that if the experimentally determined value does not match the value predicted by the standard model, it indicates non-standard model contributions. As one member of the CMS collaboration described it:

> so, all the measurements that we've done until now do not tell us anything of what would be the Higgs self-coupling. So, we don't have any reason that we are not supposed to find *new*, new physics, okay? What we measured up to now could be very well consistent – with *weird theory* and the Higgs self-coupling could be ... very different from, from the Standard Model prediction.
>
> (emphasis speaker's own)

This is one source of the potential epistemic value of this experiment: as an experimental pathway to knowledge of BSM physics. This bottom-up approach differs significantly from top-down approaches such as direct searches for SUSY particles that have the potential to find evidence for the predicted particles of certain SUSY approaches. With a bottom-up approach, the possibilities are more open, with the potential to generate heterogeneous pathways. This potential comes from the work that will be required by a result that differs from the standard model expectation: such a result will need to be accommodated into the standard model, transforming the standard model, and will thereby offer insights into BSM physics by putting constraints on the nature of the BSM contributions. The potential is, in a sense, wider in that accommodating a new result could offer evidence for one of the many predicted BSM models, but it also may require a wholly new approach for BSM physics.

An illustrative example where a similar result as to that aimed for in precision measurements occurred in 2015–2016. The ATLAS and CMS experiments found an excess above the standard model expectation that could not easily be reconciled with the standard model or any existing BSM approaches. The result was ultimately found to be a statistical fluctuation but one researcher in ATLAS described it thus:

> we had a broad signal and a broad signal to diphotons is very hard to explain. It was a really ugly thing actually, so you needed a lot of new physics to explain it. It would be actually fantastic if it would have been true: we would have had a huge amount of new physics.
>
> (quoted in Ritson 2019)

This researcher here captures the reasoning behind the value of ugliness. The uglier the result the more that will have to be done to accommodate it, and the more it will transform the understanding of BSM physics.

9.4 The Nature and Value of Ugliness

For the most part reflections on aesthetic considerations in sciences and scientific practices have focused on the nature and value of beauty and the sublime: see, for example, McAllister (1999) and the contributions to the edited volume *The Aesthetics of Science: Beauty, Imagination and Understanding* (2020). Recently, Ivanova (2020) drew on a non-factive account of understanding proposed by de Regt (2017), which treats understanding as a skill or ability to explore the epistemic value of imagery and representation. In "The Aesthetics of Scientific Experiments", Ivanova (2021) further outlined some of the ways in which beauty is understood to have value in the context of the sciences: beauty is linked to simplicity, symmetry, clarity of results, clarity of significance, results and theoretical developments that settle questions, and aptness. The measurement of the self-coupling of the Higgs is in part[5] motivated by the value and potential for an ugly experimental result, where an ugly experimental result may (a) be difficult, (b) violate symmetry principles, (c) be unclear, (d) raises questions, and/or (e) be apt. In important ways, ugliness is in opposition to characteristics linked to beauty, with aptness being the one exception to which I will return later.

First, let us unpack a little this list of characteristics linked to ugliness in the context of the measurement of the self-coupling of the Higgs, and in what sense the characteristics may be valued. It is first important to distinguish in what sense a lack of clarity should be understood here. A lack of clarity (c) here does not refer to the validity of the measurement result, but rather to how the result can be interpreted with respect to what is known (the standard model and the existing BSM models) and how it raises questions (d). There can be no value in a result that is transient or flawed, such as occurred in 2011 when the OPERA experiment at CERN announced that it had measured neutrinos travelling faster than the speed of light, a result that appeared to contradict general relativity. This result was later attributed to a loose cable responsible for connecting a GPS satellite receiver to a computer. It is important that the results obtained are valid and have a clear statistical significance.[6] In the case of the measurement of the self-coupling of the Higgs boson, it is necessary to be able to compare the experimentally determined result to the prediction determined from the standard model. This is the essential step in establishing whether the Higgs is playing the role that the standard model predicts, and therefore in determining the broader significance of the result. Where a lack of clarity can have value is with respect to the significance of the result: the result itself is robust, but the interpretation is unclear.

The notion of violating symmetry principles (b) is more intuitive. This is a characteristic we can attribute to experimental results that contradict, undermine, or challenge symmetry principles, results that offer evidence for asymmetry. This notion of violation speaks to how ugliness can be in opposition to the sublime, as violation implies defilement and destruction. To be violating is to be heretical and disrupting. Ritson (2019) explores disruption in experiments, arguing that disruption can be exploratory. Ritson offers up disruptive exploratory experimentation (DEE) as a complementary understanding of experimentation to Steinle's (1997; 2002) and Elliott's (2007) notions of exploratory experimentation (EE) and argues that

> in both cases, EE and DEE, theoretical confirmation is not sought. In the case of [EE], according to Stienle and Elliott, it is due to lack of existing theory. In the case of [DEE], it is due to an aim to deliver a result that in some way contradicts the existing theory. However, DEE is still exploratory in that a precise result is not specified or aimed for but a more general result is aimed for, one that disrupts. (p. 8)

In the case of the yet-to-be-completed Higgs boson precision measurements, we can understand experiment is exploratory in that it is designed to make possible a result that could contradict one of the many symmetry principles that underpin the standard model. A result that violates or disrupts existing symmetry principles result has the potential to drive novel and alternative epistemic strategies and principles or give an indication of how existing symmetry principles might be modified. It is perhaps too strong to say that the potential of disruption is to offer evidence of "what went wrong" given the extraordinary successes of the standard model, but instead, a violation of symmetry principles may offer guidance in the very specific "context of non-discovery" of expected extensions to the standard model, in which context theoretical and experimental physicists currently find themselves.

Difficulty (a), like simplicity, seems like an intuitively obvious characteristic; however, it is maddeningly difficult to be precise about both difficulty and simplicity. Kuhn claimed that simplicity was a characteristic of a good scientific theory and should be understood as 'bringing order to phenomena that in its absence would be individually isolated' (Kuhn 1977, 103). Much of the discussion of simplicity focuses on the ways in which simplicity is ill-suited as a determinant for theory choice given that the concept permits multiple interpretations, such as simplicity as unifying power (following Kuhn), or simplicity as parsimony, and that simplicity is not sufficiently well defined (for example, 'does simplicity concern the ontological commitments of a theory or its mathematical form?' (Bird 2022). These (theory-focused) discussions do, however, assume some virtue in simplicity.

Difficulty, understood as the antithesis of simplicity, is rarely assumed to be virtuous. The characteristic of difficulty is connected to incomprehensibility or complexity in this context (and is related to the discussion of unclear significance above). The potential of the measurement of the self-coupling of the Higgs boson is to deliver a result that both contradicts the standard model and is difficult to accommodate by the standard model. Furthermore, in contrast to a direct search that has the potential to offer evidence for specific BSM models, the measurement of the Higgs self-coupling has the potential to generate results that are not easily understood or accommodated by the many existing BSM approaches. As Kieran (1997) has noted, 'our interest in the ugly, incoherent or obscene lies in the curious violation of our social norms and conceptual framework' (p. 388). Curiosity, and the connection to difficulty, is also an interesting notion to reflect upon here: a violation of the existing conceptual framework, the standard model, can drive curiosity into considerations of how it could otherwise be. Difficulty is connected to a notion of a problem to be solved or a question to be answered, a problem or a question that may have otherwise not been raised.[7]

What is important to note is that in the context of the attempts to measure the self-coupling of the Higgs boson, ugliness has epistemic value. Kieran (1997) has explored potential value in ugliness more broadly and has argued that 'we, or at least some of us, seem to value in art images or music which are assumed to fall out-side the sphere of aesthetic value because they are ugly, grotesque or incoherent and that is their very point' (p. 386). Ugliness has the potential to be generative, however, the context is crucial, a point Kieran has also stressed: 'it is not that ugliness and incoherence are themselves intrinsically valuable in aesthetic terms, but rather that under certain circumstances they can be used to make us confront and explore certain possibilities about our own society, possibilities we would not otherwise think about' (Kieran p. 389). In the very specific historical context of non-discovery, considerations of what is apt (e) might in other contexts look unusual, challenging, or undesirable.[8]

Stanford has argued that the 'institutional apparatus of contemporary scientific inquiry fosters an exceedingly and increasingly theoretically conservative form of that inquiry' which results in 'the problem of unconceived alternatives' (Stanford 2006, 2019). The problem, Stanford argues, is that the historical record of scientific enquiry 'offers abundant evidence of the repeated failure of scientists and scientific communities to even conceive of fundamentally distinct alternatives to extant theories that were nonetheless both scientifically serious and reasonably well-confirmed by the evidence available at the time' (2019, pp. 3915–3916). Stanford further argues that the historical rise of 'Big Science' has 'served to reduce not only the incentives but also the freedom scientists have to pursue research that challenges existing theoretical orthodoxy or seeks to develop fundamental theoretical innovation' (p. 3917).

It is hard to escape the conclusion that the ATLAS and CMS experiments should be considered "Big Science": the LHC and the experiments located there are often considered to be the archetypal "big science" experiments. The sheer size, timescales, number of investigators, level of hierarchy, and international collaboration involved in the experiments dwarf most human knowledge-creating endeavours. The timescale for the attempts to measure the self-coupling of the Higgs boson is estimated to be over 12 years into the next phase of the LHC: the High Lumi Large Hadron Collider (HL-LHC) is scheduled to begin collecting data in 2029. Whilst Stanford raises the problem of unconceived alternatives as a problem for scientific realism, Stanford is also revealing the key to the epistemic value of ugliness where he highlights the difficulty, especially for "Big Science", in conceiving of fundamentally distinct alternatives.

At face value, to value a lack of understanding or a result that resists easy or initial attempts to be understood or incorporated into existing conceptual frameworks, seems paradoxical. However, the value of ugliness, the value of a lack of understanding, is that it is a path to BSM physics that has never been described or proposed or an "unconceived alternative". The problem of designing experiments that are optimised for unknown results is difficult. Feasibility considerations necessitate that experiments are designed to optimise the possibility of detection of the signatures of particles with some assumed properties. That is, the design is partly driven by what is expected. Some of this difficulty can be seen in the aforementioned example of the long-lived particles. The ATLAS and CMS detectors were designed assuming that the particles created in the collisions would decay instantaneously. Because long-lived particles decay away from the collision point, the experimental collaborations have had to devise new methods to find evidence for long-lived particles, such as dealing with unexpected background sources. However, in this example, the new methods are designed around assumed properties of long-lived particles, even if they differ from the original design assumptions. The potential of a measurement result that cannot easily be understood using the standard model is that it could provide a pathway to something never conceived of by experimental and theoretical physicists.

Of course, there is no guarantee of an ugly result in the attempts to measure the self-coupling of the Higgs boson: the measurements are yet to be completed and results are unknown. The attempts to measure the self-coupling of the Higgs in the CMS and ATLAS collaborations are a fascinating example of experimental design with the potential to provide an indirect path to the unconceived, a way to "unknown unknown" BSM physics as opposed to the many conceived alternatives, in the form of predicted BSM models. Ugliness has the potential to generate new knowledge through puzzle solving and to overcome the bias from prior beliefs, even the often-inescapable bias of having to design an experiment to search for evidence with at least some known properties.[9] Ugliness can be a way to exceed imagination – a pathway to something unconceived.

However, it is very unlikely that this is a path to overthrow the whole standard model: the expectation for an unconceived alternative is an alternative that retains most of the standard model (perhaps as a low-energy approximation). The standard model has been extraordinarily successful and any 'scientifically serious' alternative that is built from the accommodation of an ugly experimental result will also need to be 'reasonably well-confirmed by the evidence available at the time' (Stanford 2019, pp. 3915–3916). We should also be sceptical of the enduring epistemic value of ugliness. Persistent and proliferating incoherence, questions to answer, unclear significance of results, and violation of principles do not provide an ideal environment to generate evidence and secure knowledge. A lack of stability prevents access to secure handles on which to grasp and solve problems. What has been discussed here is a specific context in which it may be possible for new knowledge to be borne out of a clear articulation of a lack of understanding. Ugliness has epistemic value in this very particular context: its nature is transformative.

Thus far, I have not discussed the various claims concerning the value of beauty or elegance of certain SUSY and supersymmetric models. Claims of the value of beauty have certainly been made and also questioned, especially in popular accounts from theoretical physicists such as Brian Greene (1999) and Sabine Hossenfelder (2018). It is also far from clear what role considerations of beauty have played in the historical development of theoretical high energy physics, a very large and diverse field (Ritson 2016). I will leave this discussion to one side: my focus here has been on the ATLAS and CMS experiments and experimental design that aims to facilitate possibilities beyond the known conceived alternatives (the many existing proposed extensions to the standard model). I am not suggesting that what I have discussed here is a reaction to the "failure of a beautiful theory". Instead, I claim that the value of ugliness here is as a path to novelty: the uglier the result, the more "new physics" it might take to accommodate it. Aesthetic considerations are not truth conducive here (or truth "inconducive", as would be the analogous case). The ATLAS and CMS experiments have the potential to offer hints of, and to provide the first steps down, a pathway to somewhere not conceived of before.

Notes

1 For more on the incredible complexity and knowledge conditions for the LHC experiments, see Beauchemin (2017).
2 See, for example, (Hoyningen-Huene 1987; Leplin 1987; Schickore & Steinle 2006).
3 See, for example (ATLAS collaboration 2021).

4 For more on the current attempts to complete this measurement and the conditions, see Ritson (2021).
5 Of course, the potential and value of a measurement of the self-coupling of the Higgs boson that is consistent with the standard model expectation is also a motivating factor. Such a measurement would be a very significant result and would be an important part of the discovery claim for a standard model Higgs boson.
6 For the history of standards for discovery/observational claims in experimental particle physics, see Franklin (2013).
7 This connects to the work of Murphy in this volume on profundity, especially where Murphy draws attention to genuinely profound experiments as being those experiments that 'confound' existing theoretical approaches and 'nudge' scientific communities to develop new theories to interpret the results.
8 The importance of the context here raises questions about the future stability of aesthetic judgements. Claire Anscomb in this volume argues that ugly art can ultimately be beautiful with a change of focus as to how design achieves results.
9 I should also note that precision measurements are not the only bottom-up experimental high energy physics in the LHC experiments: the so-called 'model independent searches' also pursue this strategy. For more on these attempts, see Bechtle et al. (2022).

Bibliography

Allanach, B. (2018) Going Nowhere Fast: After the success of the Standard Model, experiments have stopped answering to grand theories. Is particle physics in crisis? *Aeon Magazine.*

ATLAS Collaboration. (2003) 'ATLAS high-level trigger, data-acquisition and controls: Technical Design Report ATLAS-TDR-016'.

ATLAS Collaboration (2021) 'Search for Displaced Leptons in √13 = TeV Pp Collisions with the ATLAS Detector', *Physical Review Letter*, 127(5), 051802

Beauchemin, P. H. (2017) 'Autopsy of Measurements with the ATLAS Detector at the LHC', *Synthese*, 194(2), 275–312.

Bechtle, P. et al. (2022) Bottoms Up: Standard Model Effective Field Theory from a Model Perspective arXiv preprint *arXiv:2201.08819.*

Bird, A. (2022) 'Thomas Kuhn', *The Stanford Encyclopedia of Philosophy* (Spring 2022 Edition). https://plato.stanford.edu/cgi-bin/encyclopedia/archinfo.cgi?entry=thomas-kuhn

CMS collaboration. (2002) 'Data Acquisition and High-Level Trigger Technical Design Report'. CERN/LHCC-26. https://cds.cern.ch/record/578006/files/cer-2336481.pdf

De Regt, H. W. (2017) *Understanding Scientific Understanding.* Oxford: Oxford University Press.

Elliott, K. C. (2007) 'Varieties of Exploratory Experimentation in Nanotoxicology', *History and Philosophy of the Life Sciences*, 29(3), 313–336.

Frank, A and Gleiser, M. (2015) 'A Crisis at the Edge of Physics', *The New York Times.*

Franklin, A. (2013) Shifting Standards: Experiments in Particle Physics in the Twentieth Century. University of Pittsburg Press https://upittpress.org/books/9780822944300/

Greene, B. (1999) *The Elegant Universe: Superstrings, Hidden Dimensions, and the Quest for the Ultimate Theory*. London: Jonathan Cape.
Hossenfelder, S. (2018) *Lost in Math: How Beauty Leads Physics Astray*. New York: Basic Civitas Books.
Hoyningen-Huene, P. (1987) 'Context of Discovery and Context of Justification', *Studies in History and Philosophy of Science*, 18, 501–515.
Ivanova, M. (2020) 'Beauty, Truth and Understanding', in *The Aesthetics of Science: Beauty, Imagination and Understanding*, Ivanova M. and French S. (eds.) New York: Routledge, pp. 86–103.
Ivanova, M. (2021) 'The Aesthetics of Scientific Experiments', *Philosophy Compass*, 16(3), 1–9.
Ivanova, M. and French, S. (eds.) (2020) *The Aesthetics of Science: Beauty, Imagination and Understanding*. New York: Routledge.
Kieran, M. (1997) 'Aesthetic Value: Beauty, Ugliness and Incoherence', *Philosophy*, 72(281), 383–399.
Kuhn, T. S. (1977) "Objectivity, Value Judgment, and Theory Choice", *The Essential Tension*. Chicago, IL: University of Chicago Press, pp. 320–339.
Leplin, J. (1987) 'The Bearing of Discovery on Justification', *Canadian Journal of Philosophy*, 17(4), 805–814.
Lykken, J. and Spiropulu, M. (2014) 'Supersymmetry and the Crisis in Physics', *Scientific American*, 310(5), 34–39.
McAllister, J. W. (1999) *Beauty and Revolution in Science*. Ithaca, NY: Cornell University Press.
Ritson, S. (2016) *The Many Dimensions of the String Theory Wars. PhD Thesis-University of Sydney.*
Ritson, S. (2019) "Probing Novelty at the LHC: Heuristic Appraisal of Disruptive Experimentation", *Studies in History and Philosophy of Science Part B: Studies in History and Philosophy of Modern Physics*, 51, 44–56.
Ritson, S. (2021) "Creativity and the Self-Coupling of the Higgs: Comparing Dual Models of Measurements at the LHC", *Synthese*, 199(5), 11887–11911.
Schickore, J. and Steinle, F. (2006) *Revisiting Discovery and Justification. Historical and Philosophical Perspectives on the Context Distinction.* Dordrecht: Springer.
Stanford, K. (2006) *Exceeding Our Grasp: Science, History, and the Problem of Unconceived Alternatives*. New York, NY: Oxford University Press.
Stanford, K. (2019) 'Unconceived Alternatives and Conservatism in Science: the Impact of Professionalization, Peer-Review, and Big Science', *Synthese*, 196, 3915–3932.
Steinle, F. (1997) 'Entering New Fields: Exploratory Uses of Experimentation', *Philosophy of Science*, 64(S4), S65–S74.
Steinle, F. (2002) 'Experiments in History and Philosophy of Science', *Perspectives on Science*, 10(4), 408–432.
Wolchover, N. (2016) 'What No New Particles Means for Physics', *Quanta Magazine*.

10 The Future Won't Be Pretty

The Nature and Value of Ugly, AI-Designed Experiments

Michael T. Stuart

Can an ugly experiment be a good experiment? Philosophers have identified many beautiful experiments and explored ways in which their beauty might be connected to their epistemic value. However, in this chapter, I seek out (and celebrate) ugly experiments. Which experiments are those? For hardcore theoreticians, perhaps *all* experiments are ugly, and only pure theory can be beautiful. For others, "ugly" experiments will be shorthand for "bad" experiments, similarly to how "beautiful" can be shorthand for "good" (Todd 2008). To avoid these ambiguities, this chapter will focus on distinctly *aesthetic* ugliness in scientific experiments to consider what, if anything, the relationship might be between this and epistemic value, and specifically, whether aesthetic ugliness can ever be epistemically good.

10.1 Introduction

There are at least two things we should be clear about when evaluating experiments. The first is to specify which *aspect* of an experiment we want to evaluate. Aspects of experiments include the instruments, the results, the objects, the hypothesis being tested, the performance, the interpretation, and the design. Following a number of authors, I will focus on evaluations of the *design* of experiments (Parsons and Rueger 2000; Murphy 2020; Ivanova 2021, 2022). This is because it is the design that defines the experiment. In many cases, we can use the same experimental design to test different hypotheses related to different objects, get different results, interpret the same results differently, and swap in different instruments. But we cannot alter the central features of an experiment's design without creating a new experiment.

The second thing to be clear about is the sense of goodness at issue. An experimental design can be *ethically* good, for example, when it does not irresponsibly risk harm, or when it aligns with the ethical values of those who stand to profit or be harmed by the outcomes. The design of an experiment can also be *epistemically* good, such that it is likely to produce a high signal-to-noise ratio, eliminates possible confounds, or is easy to

DOI: 10.4324/9781003263920-11

replicate. Finally, the design of an experiment can be *aesthetically* good, such that its design is elegant, clear, simple, symmetrical, or beautiful.

These distinctions gloss over a difficult axiological question: are ethical, epistemic, and aesthetic values merely different manifestations of a single, maximally general kind of goodness (like the Platonic form of the Good), or are they indeed fundamentally different kinds of goodness? While interesting, we will leave this aside to ask the more applied question concerning how far these different senses of goodness can come apart in the case of experimental design in science. This question is meaningful independently of whether the different kinds of goodness are manifestations of a more general good.

One way to begin answering this question is to look at examples. John Horgan (2012) collects some of 'science's ugliest experiments', including the Castle Bravo atomic bomb test in 1954, syphilis trials in Guatemala in the mid-1940s, and using electrical brain stimulation to "treat" homosexuality between 1949 and 1980. Some of these experiments were ugly in the sense of being ethically bad, while others were epistemically bad, and some were both. However, it seems that an experiment's design can be ethically good while being epistemically bad. Consider members of institutional *ethics* review boards, who do not need to care whether an experiment's design has a high likelihood of producing a result with *epistemic* value. That is not their job. And going in the other direction, we note that there are experimental designs that would accurately reveal causal dependencies in a system (this is one way to be epistemically good) while causing harm to humans. The Castle Bravo experiment might be an example of this.

Likewise, aesthetically beautiful designs can be ethically bad. For example, imagine a very clear and simple design that tests the effect of some poisonous substance on humans. Going in the other direction, we can also imagine ethically flawless experiments that are convoluted, needlessly complicated, inelegant, and so on.

This brings us to the relation between aesthetic and epistemic goods in experimental design, which is the focus of this chapter. Let's put a few options on the table. It might be that aesthetic values are completely independent of epistemic values, at least when it comes to experimental design, such that having one kind of value tells us absolutely nothing about whether the design has the other kind of value. Alternatively, one kind of value might be *sufficient* for the other, such that possessing one kind of value automatically tells us that an experiment has some value of the other kind. Another possibility is that one of these values is *necessary* for the other. For example, perhaps an experiment's design can't be beautiful if it's not likely to tell us something new about the world, or a design can't be epistemically good if it's not at least *somewhat* clear, simple, elegant, and so on. Another possibility is that there is no general connection between beauty and epistemic value, but there is/are connections between *certain* aesthetic properties and *certain* epistemic properties. For example, perhaps symmetry is tied to robustness, but not to having a good signal-to-noise ratio. A final possibility is that increasing one kind of value might generally *tend* to increase the other kind of value, without guaranteeing or requiring it.

To conserve space, we will focus only on the one-way inference from aesthetic value to epistemic value. This is perhaps the more interesting direction of inference, as it inspires strong and polarising intuitions. Consider, for example, the statement made by Roger Penrose quoted by James McAllister: 'something which looks attractive may have a better chance of being true than something which looks ugly... So often, in fact, it turns out that the more attractive possibility is the true one' (McAllister 1996, 90–91). Some will find this tantalising while others will find it simply wrong.

We will also restrict ourselves to particular and technical notions of aesthetic and epistemic values. Roughly, by epistemic value I mean something like "internal validity", that is, the ability of a design to reliably, replicably, and robustly identify the (causal) effect of one variable on (an)other(s) (Campbell 1957; Guala 2003, 2005; Hogarth 2005; Jimenez-Buedo and Miller 2010). And by aesthetic value, I mean something like a positive, pleasurable feeling resulting from first-hand (intellectual, imaginary, or perceptual) experience (Schellekens 2022). Other forms of epistemic and aesthetic value (including understanding as a kind of epistemic value and the sorts of aesthetic values more *en vogue* in modern discussions of art) will also be discussed, though these will not be the main focus. Finally, I will mostly be concerned with aesthetic and epistemic values in general rather than the properties that are referred to in justifying aesthetic and epistemic judgements (e.g., simplicity, clarity, signal-to-noise ratio, etc.) since it would be impossible in one chapter to address all the relationships between all the properties. In sum, here are the options.

1. *Independence*: the positive or negative aesthetic value of an experimental design tells us nothing about its epistemic value.
2. *Proportionality*: increasing the aesthetic value of an experiment's design generally tends to increase its epistemic value.
3. *Necessity*: if an experiment's design has epistemic value, it has aesthetic value.
4. *Sufficiency*: if an experiment's design has aesthetic value, then it has epistemic value.
5. *Necessity and sufficiency*: an experimental design has epistemic value if and only if it has aesthetic value.

Which of these is the correct way to understand the relationship between aesthetic and epistemic value? In the next section, I will present a case study that takes the last three options off the table, leaving only the first and a specific, weakened reading of the second. This may be surprising, since most of the work on the aesthetics of experiments focuses on the beautiful (not the ugly), and it is easy to get the impression from the literature that the connection is quite strong. On the other hand, some readers will already be convinced that there is no (strong) aesthetic-epistemic relationship. In that case, my contribution is to provide new evidence for that intuition.

My argument begins by noting that AI is currently designing experiments. These AI-designed experiments are here to stay, and one day they might represent the majority of experiments in science. (This is important because if AI-designed experiments were just an oddity, we would be tempted to exclude them from any discussion of the aesthetics of experiments). These experiments are ugly, and the nature of the algorithms that produce them helps to explain *both* their ugliness *and* their epistemic value.

10.2 AI-Designed Experiments

In the mid-1960s, the success of thinking machines and "analytic engines" in certain computational tasks soon led to the idea of creating a robot scientist. Herbert Simon's thinking machine, the Logic Theorist, combined with the new Information Processing Language, inspired researchers like Edward Feigenbaum to create discovery programs that were applied in the DENDRAL project, whose purpose was to automate mass spectrometry experiments for the Viking landers, which would test for amino acids on Mars (Lindsay et al. 1993; Feigenbaum 2007). This idea was scrapped because of the energy and weight constraints of extraterrestrial robotics, but it was taken up elsewhere in chemistry, for example, by a program called FAHRENHEIT whose purpose was to perform experiments, analyse data, look for errors, determine replicability, and create new theoretical hypotheses, all in the service of (among other things) detecting low-concentration ions (Żytkow 1987; Żytkow et al. 1990).

In the early 2000s, Ross King oversaw the creation of several robot scientists. These were "closed-loop" systems, in the sense that they designed, performed, and interpreted experiments, and then fed what they found back into the system to inspire new experiments, and so on until a solution was produced. The first of these, Adam (2004–2011), worked on functional genomics in yeast, performing gene knock-out experiments (King et al. 2004). Initially, it found the functions of genes that were already known. Then it discovered genes responsible for the production of important amino acids, which were not known. These discoveries were checked manually, and Adam was correct. King claims this is the first robot scientist to generate completely novel scientific knowledge. His next project, Eve (2008–2020), focused on neglected tropical diseases, identifying compounds that might be useful in producing cheap and effective treatments (Williams et al. 2015). Eve was also recently used to test the replicability of studies in cancer research (Roper et al. 2022). A new project, Genesis, is now deployed in the context of systems biology, and it is advertised as being able to perform up to 10,000 closed-loop cycles of experiments at a time.

One admitted shortcoming of these prototype projects is that human intervention is often required, and the experiments are relatively conservative variants of existing designs. However, recent work by Mario Krenn and colleagues has produced an algorithm ("MELVIN") that designs experiments that no one has ever thought of before. More specifically, Krenn's team could not come up with a viable design for a certain type of experiment, and the algorithm succeeded where the humans couldn't in suggesting a way to arrange laboratory equipment such that certain quantum phenomena could be produced and manipulated in the lab. As Krenn et al. point out, AI is already being used in a number of ways relating to the design of experiments in accelerator physics, plasma physics, nanophotonics, quantum circuits, and superconduction (Krenn et al. 2020), but using AI to design new experiments from scratch is only now gaining ground.

Krenn's lab wanted to produce a certain type of quantum entangled system. The most common type of entanglement (the "qubit") obtains between two particles and two states (e.g., spin up and spin down). Higher-dimensional entanglement occurs when we have more than two particles and/or more than two states. This kind of entanglement is potentially very important for quantum computing, as it has greater information-carrying capacity. Trying to produce and manipulate certain types of higher-dimensional entanglement had proven 'difficult' (Erhard et al. 2020): Krenn reports trying with colleagues for several weeks without success to produce a particular form of multiparticle entangled state (Krenn 2021). The algorithm MELVIN was developed, and it produced an experiment that was 'the first genuine high-dimensional multiparticle entanglement' (ibid), using crystals to produce photon pairs. This discovery (Erhard et al. 2018) enabled independent manipulation of the quantum states, invented new tools for quantum experiments, and has now been expanded for use with non-photonic systems.

Roughly, the algorithm possesses a 'toolbox' filled with representations of optical laboratory components, which it combines randomly. The combinations are tested for viability, and those that pass are then tested in more detail. Even if a set-up is not viable, it might produce a new type of tool, which is then added to the toolbox. The final set-up that MELVIN recommended required the use of tools that it conceptualized on its own. Experiments are represented mathematically as graphs, with vertices representing photon paths and edges representing amplitude by their weights and quantum properties by their colours. The successful outcome is then simplified by an XAI algorithm that eliminates redundant elements so that humans can understand it well enough to implement it.

AI-designed experiments still make up a minority of all scientific experiments performed, but there are reasons to think that this will change. Frank Wilczek, who won the Nobel Prize in physics in 2004, has claimed that in 100 years the best physicist might be a machine (2016). The Nobel Turing Challenge is a competition to produce a Nobel-winning AI scientist by 2050 (Kitano 2021). Ross King argues that we should expect the use of AI to expand in science because it is reliable when it comes to memory and mathematics, it can process more data more quickly than humans can, and it can find patterns that are invisible to humans. He points out that AI is very good at anything that can be gamified, and quite a bit of science can be gamified. AI tends to be cheaper, faster, more accurate, more detailed, and it can be scaled up and controlled remotely (King 2021).

We therefore have good reason to believe that "experimental design" should not only refer to human-designed experiments, but experiments designed either by humans, by human-AI teams, or by AI themselves.

10.3 The Good, the Bad, and the Ugly

In *Scientific American*, Aephraim Steinberg comments that the experiment designed by MELVIN is 'a *gorgeous* first example of the kind of new explorations these thinking machines can take us on' (Ananthaswamy 2021; emphasis added). Steinberg did not work on these experiments, but he uses AI to design experiments of his own. I reached out to Steinberg, and the quotations that follow come from our conversation, which took place over a few weeks in 2021.

So, what did Steinberg mean when he said that the experiment was gorgeous? He recognised that scientists 'use anthropomorphic terms like "beautiful" and even "sexy" in vague and ill-defined ways all the time', but this was not what he was doing in this case. 'There was something... lip-smacking, entrancing...about this implementation, particularly to experimentalists. Not that it's the most important, the most creative, or the most elegant experiment ever, but...to my eye, at least, there's a 'rhythm' and 'cadence' to the series of interferometers in their experimental setup which indeed looks almost poetic or artistic'. He goes on, 'I think it was the combination of the experimental prowess and elegance, with the fact of implementing the 'clever' scheme from the AI-invented protocol which pushed me to describe it the way I did...I liked this example because I thought they really did demonstrate that they could do something that as far as I know, no humans had figured out how to do before'. In sum, Steinberg's notion of a beautiful experiment involves newness, surprisingness, simplicity of design (i.e., elegance as opposed to arcaneness), looking good (either in the pdf or physically on the tabletop), and having good rhythm or cadence in the performance (whether that performance is imagined or physical).

This matches quite closely what philosophers have said about beautiful experiments in the past. Glenn Parsons and Alexander Rueger focus on the aptness of a design, the simplicity of its steps, and the imagination of its designer (2000). Milena Ivanova focuses on the elegance, simplicity, and clarity of a design (Ivanova 2022). Steinberg agrees with these and adds rhythm or cadence, which is a nice addition to the set. Grouping together these properties, we have something consistent with what Elisabeth Schellekens calls the 'standard conception' of aesthetic value. For Schellekens, the standard conception is the classic, traditional one, which develops from the following two axioms: (a) aesthetic experience is grounded in first-hand perception, and (b) aesthetic experience is characterised by pleasure (Schellekens 2022, 125). In our case, we should expand the first axiom so that it refers to *experience* rather than perception, since experimental designs are sometimes imagined or contemplated rather than literally perceived (call this expanded version the "standard+" conception of aesthetic value). Then, a beautiful experiment is one whose design is experienced first-hand, where that experience is a pleasurable one. That pleasure will be tied to experiences of particular aspects of the design, like its aptness, clarity, rhythm, etc.

As Schellekens points out, the history of aesthetics is a history of expanding this standard conception to include many different kinds, sources, and media for aesthetic value. But these various kinds of modern and post-modern aesthetic values are not what is at issue in discussions of the aesthetics of science. Scientists and philosophers are not concerned with whether an experiment can have aesthetic value in the same sense as the art of Duchamp, Picasso, Rothko, Pollock, John Cage, or Ai Weiwei. Perhaps they should be, but they're not. Therefore, in what follows, by "beauty" we will refer to the most general standard+ aesthetic value and by "ugly" we will refer to the lack of such value.[1]

On the standard+ conception of aesthetic value, is MELVIN's experiment beautiful? It might seem so, from what Steinberg said above. However, in an interesting turn of conversation that inspired this chapter, Steinberg focused directly on the aesthetic properties of the experiment and in so doing, began to suggest a different view. He went on, 'the part of the experiment the AI designed is probably less elegant than many of the simple setups humans have designed for various experiments along the way'. At this point, he begins to separate the contribution of the AI from the contribution of the humans. 'When I talk about the "experiment," I mean this chip the group put together, and… the computer didn't tell them how to do that'. This is important, as it recognises a general feature of many modern AI designs. Whether AI is used in designing an experiment, a chair, a car, a building, or something else, many of the features that humans find aesthetically pleasing are those added by the humans who chose, developed, and implemented the design.

Further, 'if the solution is elegant, it's not so hard for people to stumble upon, while if it's arcane, it may be harder for us to discover'. In other words, when the experimental design is 'nice and elegant', we don't need the AI, because 'people would have gotten around to inventing it if they'd had much reason to'. This is important because it reminds us that the experiments we need AIs to design won't usually be ones with clear, simple, elegant solutions. They'll be the arcane, ugly ones.

Steinberg portrays the general situation like this,

> With genetic algorithms, or other optimization techniques, they've been able to develop remarkable pulse sequences for controlling quantum systems. And usually, they just look like a godawful mess that happens to work... In some cases, you could squint your eyes and say 'oh! it's just trying to do [some technique we already knew about],' and it would all make sense. And it would turn out that all the 'mess' was just 'noise,' and if a human had designed it, it would have been 'prettier' (simpler, cleaner), but worked just as well. I believe by now there are cases where this isn't true, and we don't know of 'clean' optimal solutions, so have only the computers.

To illustrate this point, Steinberg referred to some work of his own (Figure 10.1).

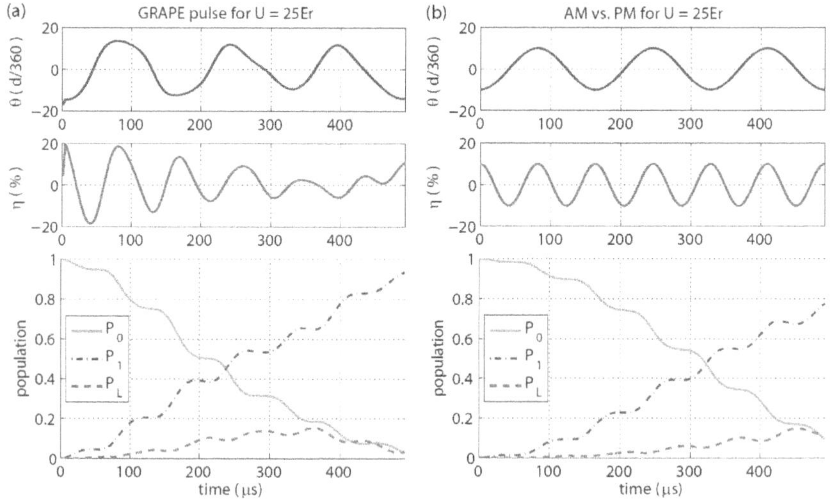

Figure 10.1 An example of an "ugly" use of AI in a quantum experiment (reproduced with permission from Zhuang et al. 2013). The original caption reads: 'GRAPE pulse and a similar AM vs. PM pulse. (a) GRAPE pulse and the corresponding simulation results with the lattice Hamiltonian without gravity. (b) A similar AM vs. PM pulse has the same pulse duration and the variance as the GRAPE pulse and the corresponding simulation results with the lattice Hamiltonian without gravity'.

Steinberg explains that the figure

> Shows two clean sinusoids (in the upper right) which implement a technique that makes physical sense to us, and which we intentionally tried on our atoms; 'GRAPE' (which is a computer optimisation technique) invented its own optimum, shown on the upper left, and you can see that it seems to have more or less discovered the same idea. But if you look at the bottom panels, which show (in simulations) the effectiveness of the techniques (which are meant to bring the line for P_1 as high as possible by the end of the pulse, and the other two [P_0 and P_L] as low as possible), it looks like the GRAPE pulse – at first glance just a 'messier' version of our designed pulse – actually does better. And we don't know anything about why this 'uglier' pulse works better. We also don't know how much you could 'clean it up' and still have it work well.

In the end, experiencing MELVIN's experimental design (on the page or in the imagination) caused pleasure, which earned it the label 'gorgeous' on the standard+ conception of aesthetic value. But on closer inspection, when we isolate the aspects of the design that were created by the AI before it had been cleaned up and implemented by the humans, it is more likely to look like a 'godawful mess'. And this will be true for experiments designed by AI in general. It is not impossible that we could design algorithms that produce beautiful experiments in the standard+ sense. However, the point is that scientists are not motivated to do this. Why would they be? In science, we often need a way to control or manipulate systems that have a great deal of variables and/or display some chaotic behaviour, and it is in these contexts that AI will be most valuable. In such contexts, we should not expect AI designs to be beautiful, that is, simple, clear, elegant, and with a satisfying rhythm or cadence. This is because AI algorithms typically employ deep neural networks and machine learning, which consist in brute force exploration of all the options in a preselected set (or set of sets) of potential solutions in a problem space, or they alter node weights (i.e., numbers by which to multiply numerical inputs) that were randomly assigned and then adjusted until a certain level of performance is reached, defined in terms of finding a desired local minimum (i.e., a "solution"). Their power comes almost entirely from calculation speed. There is nothing in the code that tells the AI to prefer solutions that produce pleasure in humans, and that is a good thing, because it would limit the set of solutions the AI would consider. We use AI precisely because it is not limited in the same ways that humans are. It finds solutions that aren't the ones that we would look for because its "processing fluency" is completely different from ours. Humans find pleasure in simple, clear, elegant designs, because that is what we like to process, because it is easy to process. AI transcends those limitations. This explains why we are seeing AI used in certain parts of science and not others. Where there are many variables, heaps of data, and we need subtle, complex solutions, scientists will increasingly turn to AI.

Note that the nature and value of AI designs is similar, both aesthetically and epistemically, to AI proofs in mathematics. Viktor Blåsjö argues that 'an ugly proof resorts to computations, algorithms, symbolic manipulation, ad hoc steps, trial-and-error, enumeration of cases, and various other forms of technicalities. The mind can neither predict the course nor grasp the whole; it is forced to cope with extra-cognitive contingencies' (2018). Similarly, Alan Cain writes about computer-assisted proofs in which 'readers cannot see why this construction or this calculation or that definition is being carried out; they cannot perceive a reason for it that is internal to the proof...the deus is a cataract that their intuition cannot easily navigate' (2010). Cain is inspired by G. H. Hardy's view of mathematical beauty as unexpected inevitability, which a computer-assisted proof does not provide. Ulianov Montano takes this view further by comparing mathematical proofs to narratives, which should have a satisfying plot to be beautiful. Montano claims that 'the experience [of studying a computer-assisted proof] has deforming narrative gaps. A computer-assisted proof shall always give us an incomplete experience, something we cannot fully appreciate, despite the fact that the proof is a perfectly acceptable and widespread method. Thus, it is not very plausible that we shall come to regard computer-assisted proofs as beautiful' (2014, 203). These quotations all have in common the idea that the products of digital methods in mathematics are ugly, and for the same reason: their outputs are ad hoc, they use brute force, and they are disfluent to the human mind. But as Montano notes, these methods are 'perfectly acceptable' and becoming widespread. In the case of the computer proof of the four-colour theorem, Montano argues that it was never really questioned by mathematicians *as a proof*. 'The issue here is not whether the proof is genuine or not, but rather that it is ugly. The ugliness of the proof is thus not explained by mathematicians' epistemic or other technical concerns' (37). The ugliness we experience is similar to watching a movie with a complicated plot that suddenly stops and reports that the main characters lived happily ever after, while refusing to explain anything more. 'The assistance of the computer impairs, or rather wrecks, the proof's storytelling' (2014, 37). A similar claim can be made concerning AI-created experimental designs: what is expected but missing are the rhythm, cadence, narrative arc, underlying coherence, and the feeling that you are in the competent hands of an expert storyteller. All of this contributes to a feeling of ugliness.

I conclude that AI-designed experiments can be both ugly and epistemically good (again, where "epistemically good" refers to something like internal validity). This ugliness is characteristic of AI use in science. Given that AI plays ever greater roles in science, we can expect more ugly experimental designs in science. Of course, what makes AI-designed experiments ugly is nothing new: many experiments in science can be found that will be arcane and without a satisfying rhythm in very similar ways. But the case with AI experiments is especially clear, so I will continue to draw upon it in what follows.

10.4 When Uglier Is Better

As a reminder, here are the options we outlined at the beginning for possible relations between aesthetic and epistemic value.

1. *Independence*: the positive or negative aesthetic value of an experimental design tells us nothing about its epistemic value.
2. *Proportionality*: increasing the aesthetic value of an experiment's design generally tends to increase its epistemic value.
3. *Necessity*: if an experiment's design has epistemic value, it has aesthetic value.
4. *Sufficiency*: if an experiment's design has aesthetic value, then it has epistemic value.
5. *Necessity and sufficiency*: an experimental design has epistemic value if and only if it has aesthetic value.

We don't need to say much about sufficiency, because no one claims that simplicity or clarity (etc.) on their own guarantee epistemic value. But there are good reasons to think that beauty is *necessary* for epistemic value. Let's consider two arguments for this claim. First, we might think that beauty is necessary for epistemic value because without beauty, scientists would have no motivation to do science, and without science, we would lose all the epistemic value that comes with science. Indeed, we know that scientists *do* find beauty motivating, especially in their search for fundamental theories, even if this sometimes holds science back (Hossenfelder 2018; Vaidyanathan and Jacobi 2022). A version of this idea is expressed by Poincaré's famous quotation: 'The scientist does not study nature because it is useful to do so. He studies it because he takes pleasure in it, and he takes pleasure in it because it is beautiful. If nature were not beautiful it would not be worth knowing, and life would not be worth living' (Poincaré 1914/2003, 22). Poincaré has a particular "intellectual" notion of beauty in mind, but however broad or narrow that conception is, few would agree that the lives of those without such experiences are not worth living. A more moderate form of this argument would be that some beauty is required for science to be the kind of thing that humans pursue. Such a view is perhaps suggested by Schellekens (2022), Ivanova (2020), and Breitenbach (2020), though I do not want to claim that any of these authors really supports a strong necessitarian view like this.

There are several ways we might respond. First, beauty is motivating, but it is not the only kind of value that motivates. The Work and Well-Being in Science project finds that while 62% of scientists were motivated to become a scientist due to considerations of beauty, this leaves 38% who were not (Vaidyanathan and Jacobi 2022). Besides beauty, scientists are motivated by the ethical good that they can achieve through science, as well as epistemic goods, like truth, knowledge, and understanding. So beauty is not necessary to motivate science. Another reply would be to consider closed-loop AI scientists, as described above, which design, perform, and interpret their own experiments. In this case, the experimental designs need not have any amount of beauty in the standard+ sense for the AI to remain "motivated". All we require is a power outlet. Thus, beauty is not necessary for epistemic value, either for humans or machines.

A second argument for the necessity claim would be that some amount of beauty is necessary for the experiment's design to be understandable, and this is one kind of epistemic value that experimental designs can have. This line of thinking is inspired by arguments made by Elgin (2020) and Ivanova (2020). For Catherine Elgin, aesthetic value can act as a gatekeeper, telling us what counts as acceptable in science. For Ivanova, our preference for beautiful things is 'deeply ingrained', such that 'taking aesthetic values in science as conditions of our cognitive makeup reflecting our intellectual interest and capacities explains why aesthetic values persist even when the best theories do not seem to quite fit our aesthetic requirements' (2020). Again, neither author states that they are arguing for the necessity of aesthetic value for epistemic value, but we can draw inspiration from them to construct such an argument.

There are several ways to respond. One is to point out that understandability is only necessary if humans are always needed to implement experimental designs. If we are talking about a closed-loop AI scientist, the kind of simplicity and clarity required for human understandability or epistemic gatekeeping is not necessary. Thus, even if some beauty is required for *human* understandability, human understandability is not required for experimental designs to have epistemic value in general, because those designs that are not understandable to humans can still be understood by machines, no matter how ugly they are. Another way to reply would be to deny that beauty is required for human understandability. After all, we can understand complicated, ugly things: we simply don't enjoy it. Finally, we might also argue that human understandability isn't the main epistemic good that scientists want experimental designs to have. In this chapter, I have focused on internal validity, which is a measure of how well an experiment identifies the nature and size of the causal effects of one variable on (an)other(s). Signs of internal validity might include a high signal-to-noise ratio, robustness to perturbation, proper isolation of the experimental system, and an established theory of the instruments used. A scientist might be able to establish that all of these are possessed by a certain experimental design, even for an experiment that is 'more like a Rube Goldberg machine' (Elgin 2020).

Pending new arguments, we can reject options 3–5 from the above list and turn to options 1 and 2, starting with 2. The claim there is that increasing aesthetic value generally *tends* to increase epistemic value. Some of the arguments presented above for the necessity claim can be weakened to support this position. For example, increasing aesthetic value generally tends to increase the motivation of scientists, which tends to increase epistemic value in science. Or, increasing aesthetic value tends to increase human understandability, which tends to increase epistemic value. In reply to each, it is important to keep in mind that the claim should not be about tending to increase epistemic value in general, but about tending to increase the epistemic value of experimental designs. (We could have made the same move in reply to the necessity version of the argument above, but it wasn't necessary.) Suppose it was part of an experiment's design to play music in the laboratory while performing the experiment. We might find that this made scientists more motivated and that might lead to more epistemic value. But it would not increase the epistemic value – at least in the sense of internal validity – of that particular experiment's design. Likewise, it might be the case that a simpler design is easier for a scientist to understand, and therefore, it might pass the gate of acceptability and scientists can easily grasp how the experiment works. But these on their own don't say much about the likeliness of that experiment to correctly identify the causal structure of a system.

Thus, we are left with the first option (independence) and a weakened version of the second option (proportionality). The second option is weakened in the sense that it would hold only for those experiments implemented by humans, and only in terms of the understandability of the design, not internal validity. With respect to the first option, things are still unclear. The reason that AI-designed experiments are ugly *and* epistemically good is due to a common cause: the nature of the algorithms involved. In this sense at least, aesthetic and epistemic value do not seem to be fully independent. But on the other hand, an experiment can be ugly and good, ugly and bad, beautiful and good, or beautiful and bad. In this (different, and perhaps more general) sense, the two different kinds of values are independent because the presence of one does not on its own tell us anything about the presence of the other.

10.5 Objections

Someone might object that AI can't "design" experiments because AI can't *do* anything, in the properly intentional sense required for human action. This might be because AI is not conscious, or because it is not connected in the right way to the objects of its "cognition". To such an objection, we can simply remove all references to AI as an active designer, and allow that AI is merely a tool, like a hammer or a paintbrush. On this presentation, we claim that when *humans* use AI to design experiments, the designs are ugly. The rest of the arguments go through as before, since we have identified an important set of experiments that are both ugly and epistemically good.[2]

Another kind of objection is really a lament: if we are headed into a future in which science has only (or mostly) ugly experiments, the world would lose an important source of beauty, and it is our duty to prevent this. To argue this way accepts the conclusion of this chapter but rejects it as repugnant. I can sympathise: a future in which all scientific experimentation was planned and carried out by AI seems somehow sad. But in reply, we can point out that such a future, however unlikely, could trade one source of beauty for others, and it is not clear that aesthetic value would decrease overall. Scientists who find aspects of experimental design work beautiful would have to move on to other pursuits; however, the new pursuits might allow for just as much beauty to be produced. And overall, a future with massively automated science might be one in which more diseases are cured, climate change is better addressed, energy and food needs are met, and so on, such that humans could lead longer, happier lives full of the production and appreciation of beauty.

Another objection would refer to human aesthetic adaptation, a feature of history fruitfully discussed by McAllister (1996). The idea is that human aesthetic values have evolved over time, and they will continue to evolve, such that the more useful AI-designed experiments come to be, the more beautiful scientists will come to find them. This accepts the premises of the argument but denies the conclusion. That is, it allows that AI-designed experiments are *currently* ugly, but claims that humans will eventually acclimatize. After all, there was more than a century of baroque art and architecture, why not again? While human aesthetic values will certainly adapt to a world increasingly suffused with AI, I think it is unlikely that future humans could change (or indeed *reverse*) their deeply ingrained preferences for simplicity, symmetry, elegance, clarity, and so on, in the way required to appreciate AI designs as beautiful in the standard+ sense. Even the most baroque art is still beautiful in that sense. Such art demands more attention to the details, and this can be challenging on first exposure. But attention reveals details that are themselves elegant, symmetrical, and so on, and the overall result is clearly pleasurable. This is not the case for designs that make no attempt whatsoever to produce something clear, simple, symmetrical, rhythmic, or united by a satisfying narrative structure. It seems unlikely that humans could adapt in the foreseeable future to find designs without such properties pleasurable.

A counterargument could be that human aesthetic values have already moved on far beyond the baroque to embrace all the diversity of contemporary art, much of which already resembles AI outputs, and that art has aesthetic value. This must be admitted. But the point of the present chapter is to challenge the claim that AI does or will design experiments that have aesthetic value *in the standard+ sense*. That is the sense with which philosophers and scientists are generally concerned, and it is not the sense with which today's artists, art critics, and aficionados are concerned. Perhaps the aesthetics of science needs an update, and we should be looking at the epistemological advantages of modernist and postmodernist aesthetic value. But that is another topic.

Finally, one might argue that if an AI were advanced enough, it might come to prefer certain strategies or develop little traditions, and experience a sense of rhythm and momentum that it could celebrate in its work. In other words, if it were a "live creature" in John Dewey's sense, then AI-designed experiments might be beautiful to the AIs themselves. We are now firmly in the realm of science fiction. But if such a future obtained, we could accept the premises but reject the conclusion of this chapter by allowing that AI-designed experiments are ugly for humans, while insisting that they have aesthetic value for *someone*, and that this aesthetic value might be connected to their epistemic value. In reply, note that AI is so useful because we can program the means to identify solution states without including aesthetic preferences. And this is one important reason that AI is able to find solutions that appear so "alien" to us (see Halina 2021). If we want AI to continue to produce useful alien solutions, either we must prevent AI from developing aesthetic values, or we must continuously build new AIs that have entirely different sets of aesthetic values from their predecessors. The second option is costly and comes with no obvious advantage, so scientists will be motivated to continue with the status quo. And in that case, AI designs will continue to be ugly to everyone who is capable of aesthetic displeasure.

10.6 Conclusion

The design of an experiment can have at least three kinds of value: ethical, epistemic, and aesthetic. These kinds of value can come apart. This chapter concerned the strength of the relationship between aesthetic and epistemic value in experimental designs: are these kinds of values independent, or does the first generally tend to increase the second, or is the first necessary or sufficient (or both) for the second? To investigate, we examined a case study of an AI-designed experiment, demonstrating that AI-designed experiments can be ugly and epistemically valuable at the same time. We can expect the use of AI in experimental design to continue to expand, and therefore this kind of ugliness should not be dismissed. Its existence puts pressure on any strong relationship postulated between aesthetic and epistemic value, since ugliness is not only acceptable in science, it can be epistemically desirable. I conclude that beauty is neither necessary nor sufficient for epistemic value, and increasing beauty will not generally tend to increase epistemic value. Whether aesthetic and epistemic values are wholly independent is an open question.

In terms of experimental design, the future might not be pretty. Nevertheless, we can expect this to help, not hinder, the achievement of science's main epistemic goals.

Acknowledgements

I would like to thank Milena Ivanova and Alice Murphy for their kind invitation to contribute to this volume, for many stimulating conversations before and after that invitation, as well as for extremely helpful comments on a draft of the chapter. I would also like to thank Milena for organising two workshops leading up to this volume, sponsored by the British Society of Aesthetics. Thanks also to the audiences at those workshops, as well as audiences at the University of York and Aarhus University, and finally, thanks to Aephraim Steinberg for the exciting conversation that planted the seed for this chapter.

Notes

1 We could also characterise ugliness not as a lack of beauty, but, following a tradition that started with Rosenkranz (1853), as its own kind of positive property. Doing so should not affect the following arguments.

2 Of course, hammers and paintbrushes change over time, and some are adorned and beautiful. But the point is that the basic functions of these tools haven't changed. Equally, user interfaces for AI algorithms in science will become easier to work with, and perhaps one day they will be beautiful. But unless the fundamental structure of AI changes, the processes and output of AI will remain ugly.

Bibliography

Ananthaswamy, A. (2021) 'AI Designs Quantum Physics Experiments beyond What Any Human Has Conceived', *Scientific American*. https://www.scientificamerican.com/article/ai-designs-quantum-physics-experiments-beyond-what-any-human-has-conceived/ (Accessed: 28 November 2022).

Blåsjö, V. (2018) 'Mathematicians versus Philosophers in Recent Work on Mathematical Beauty', *Journal of Humanistic Mathematics*, 8(1), 414–431. https://doi.org/10.5642/jhummath.201801.20.

Breitenbach, A. (2020) 'One Imagination in Experiences of Beauty and Achievements of Understanding', *The British Journal of Aesthetics*, 60(1), 71–88. https://doi.org/10.1093/aesthj/ayz048.

Cain, A. J. (2010) 'Deus Ex Machina and the Aesthetics of Proof', *The Mathematical Intelligencer*, 32(3), 7–11. https://doi.org/10.1007/s00283-010-9141-z.

Campbell, D. T. (1957) 'Factors Relevant to the Validity of Experiments in Social settings', *Psychological Bulletin*, 54(4). https://doi.org/10.1037/h0040950.

Elgin, C. Z. (2020) 'Epistemic Gatekeepers: The Role of Aesthetic Factors in Science', *The Aesthetics of Science*. London: Routledge.
Erhard, M. et al. (2018) 'Experimental Greenberger–Horne–Zeilinger Entanglement beyond Qubits', *Nature Photonics*, 12(12), 759–764. https://doi.org/10.1038/s41566-018-0257-6.
Erhard, M., Krenn, M. and Zeilinger, A. (2020) 'Advances in High-Dimensional Quantum Entanglement', *Nature Reviews Physics*, 2(7), 365–381. https://doi.org/10.1038/s42254-020-0193-5.
Feigenbaum, E. (2007) 'Oral History of Edward Feigenbaum'. http://archive.computerhistory.org/resources/access/text/2013/05/102702002-05-01-acc.pdf.
Guala, F. (2003) 'Experimental Localism and External Validity', *Philosophy of Science*, 70(5), 1195–1205. https://doi.org/10.1086/377400.
Guala, F. (ed.) (2005) 'External Validity', in *The Methodology of Experimental Economics*. Cambridge: Cambridge University Press, pp. 141–160. https://doi.org/10.1017/CBO9780511614651.008.
Halina, M. (2021) 'Insightful Artificial Intelligence', *Mind & Language*, 36(2), 315–329. https://doi.org/10.1111/mila.12321.
Hogarth, R. M. (2005) 'The Challenge of Representative Design in Psychology and Economics', *Journal of Economic Methodology*, 12(2), 253–263. https://doi.org/10.1080/13501780500086172.
Horgan, J. (2012) 'What Are Science's Ugliest Experiments?', *Scientific American Blog Network*. https://blogs.scientificamerican.com/cross-check/what-are-sciences-ugliest-experiments/ (Accessed: 28 November 2022).
Hossenfelder, S. (2018) *Lost in Math: How Beauty Leads Physics Astray*. New York, NY: Basic Books. https://www.basicbooks.com/titles/sabine-hossenfelder/lost-in-math/9781541646766/ (Accessed: 28 November 2022).
Ivanova, M. (2020) 'Beauty, Truth and Understanding', *The Aesthetics of Science: Beauty, Imagination and Understanding*. London: Routledge. https://www.routledge.com/The-Aesthetics-of-Science-Beauty-Imagination-and-Understanding/Ivanova-French/p/book/9781032337180 (Accessed: 28 November 2022).
Ivanova, M. (2021) 'The Aesthetics of Scientific Experiments', *Philosophy Compass*, 16(3), e12730. https://doi.org/10.1111/phc3.12730.
Ivanova, M. (2022) 'What Is a Beautiful Experiment?', *Erkenntnis* [Preprint]. https://doi.org/10.1007/s10670-021-00509-3.
Jimenez-Buedo, M. and Miller, L. M. (2010) 'Why a Trade-Off? The Relationship between the External and Internal Validity of Experiments', *THEORIA*, 25(3), 301–321. https://doi.org/10.1387/theoria.779.
King, R. (2021) 'The Automation of Science', *Workshop on Artificial Scientific Discovery 2021*, Max Plank Institute for the Science of Light. https://mpl.mpg.de/divisions/marquardt-division/workshops/2021-artificial-scientific-discovery/ (Accessed: 28 November 2022).
King, R. D. et al. (2004) 'Functional Genomic Hypothesis Generation and Experimentation by a Robot Scientist', *Nature*, 427(6971), 247–252. https://doi.org/10.1038/nature02236.
Kitano, H. (2021) 'Nobel Turing Challenge: Creating the Engine for Scientific Discovery', *Npj Systems Biology and Applications*, 7(1), 1–12. https://doi.org/10.1038/s41540-021-00189-3.

Krenn, M. (2021) 'A Journey from Computer-Designed Quantum Experiments to Computer-Inspired Scientific Understanding', Workshop on Artificial Scientific Discovery 2021, Max Plank Institute for the Science of Light. https://mpl.mpg.de/divisions/marquardt-division/workshops/2021-artificial-scientific-discovery/ (Accessed: 28 November 2022).

Krenn, M., Erhard, M. and Zeilinger, A. (2020) 'Computer-Inspired Quantum Experiments', *Nature Reviews Physics*, 2(11), 649–661. https://doi.org/10.1038/s42254-020-0230-4.

Lindsay, R. K., Buchanan, B. G. and Feigenbaum, E. A. (1993) 'DENDRAL: A Case Study of the First Expert System for Scientific Hypothesis formation', *Artificial Intelligence*, 61(2), 209–261. https://doi.org/10.1016/0004-3702(93)90068-M.

McAllister, J. W. (1996) *Beauty and Revolution in Science*. Cornell University Press. https://www.jstor.org/stable/10.7591/j.ctv5rf5v0 (Accessed: 28 November 2022).

Montano, U. (2014) *Explaining Beauty in Mathematics: An Aesthetic Theory of Mathematics*. London: Springer (Synthese library). https://www.academia.edu/25566436/Introduction_to_a_Naturalistic_Aesthetic_Theory (Accessed: 29 November 2022).

Murphy, A. (2020) 'The Aesthetic and Literary Qualities of Scientific Thought Experiments', in M. Ivanova and S. French (eds.) *The Aesthetics of Science: Beauty, Imagination and Understanding*. New York: Routledge, pp. 146–167.

Parsons, G. and Rueger, A. (2000) 'The Epistemic Significance of Appreciating Experiments Aesthetically', *The British Journal of Aesthetics*, 40(4), 407–423. https://doi.org/10.1093/bjaesthetics/40.4.407.

Poincaré, H. (1914) *Science and Method*. London: Dover.

Roper, K. et al. (2022) 'Testing the Reproducibility and Robustness of the Cancer Biology Literature by Robot', *Journal of the Royal Society Interface*, 19(189), 20210821. https://doi.org/10.1098/rsif.2021.0821.

Rosenkranz, K. (1853) *Aesthetics of Ugliness*. Translated by A. Pop and M. Widrich. Bloomsbury. https://www.bloomsbury.com/us/aesthetics-of-ugliness-9781472568878/ (Accessed: 30 November 2022).

Schellekens, E. (2022) 'Aesthetic Experience and Intellectual Pursuits', *Aristotelian Society Supplementary Volume*, 96(1), 123–146.

Stuart, M. T. (2018) 'How Thought Experiments Increase Understanding', *The Routledge Companion to Thought Experiments*. Routledge. https://doi.org/10.4324/9781315175027.ch30.

Stuart, M. T. (2022) 'Sharpening the Tools of Imagination', *Synthese*, 200(6), 451. https://doi.org/10.1007/s11229-022-03939-w.

Todd, C. S. (2008) 'Unmasking the Truth Beneath the Beauty: Why the Supposed Aesthetic Judgements Made in Science May Not Be Aesthetic at All', *International Studies in the Philosophy of Science*, 22(1), 61–79. https://doi.org/10.1080/02698590802280910.

Vaidyanathan, B. and Jacobi, C. (2022) *The Role of Aesthetics in Science: Survery Findings from Work and Well-Being in Science: An International Study*. Department of Sociology, The Catholic University of America, p. 10. https://workandwellbeingstudy.com/wp-content/uploads/2022/05/vaid-aesthetic-fact-sheet.pdf (Accessed: 30 November 2022).

Wilczek, F. (2016) 'Physics in 100 years', *Physics Today*, 69(4), 32–39. https://doi.org/10.1063/PT.3.3137.

Williams, K. et al. (2015) 'Cheaper Faster Drug Development Validated by the Repositioning of Drugs against Neglected Tropical diseases', *Journal of The Royal Society Interface*, 12(104), 20141289. https://doi.org/10.1098/rsif.2014.1289.

Zhuang, C. et al. (2013) 'Coherent Control of Population Transfer between Vibrational States in an Optical Lattice via Two-Path Quantum Interference', *Physical Review Letters*, 111(23), 233002. https://doi.org/10.1103/PhysRevLett.111.233002.

Żytkow, J. M. (1987) 'Combining Many Searches in the FAHRENHEIT Discovery System', in P. Langley (ed.) *Proceedings of the Fourth International Workshop on MACHINE LEARNING*. Morgan Kaufmann, pp. 281–287. https://doi.org/10.1016/B978-0-934613-41-5.50032-5.

Żytkow, J. M., Zhu, J. and Hussam, A. (1990) 'Automated Discovery in a Chemistry Laboratory', in *Proceedings of the eighth National conference on Artificial intelligence - Volume 2*. Boston, MA: AAAI Press (AAAI'90), pp. 889–894.

Name Index

Subject Index

For Product Safety Concerns and Information please contact our EU
representative GPSR@taylorandfrancis.com
Taylor & Francis Verlag GmbH, Kaufingerstraße 24, 80331 München, Germany

www.ingramcontent.com/pod-product-compliance
Ingram Content Group UK Ltd.
Pitfield, Milton Keynes, MK11 3LW, UK
UKHW022029061126
477531UK00003B/108

* 9 7 8 1 0 3 2 2 0 3 8 8 1 *